Metric Survey Specifications for Cultural Heritage

By David Andrews, Jon Bedford and Paul Bryan

 Historic England

Published by Historic England, The Engine House, Fire Fly Avenue, Swindon SN2 2EH
www.HistoricEngland.org.uk

Historic England is a Government service championing England's heritage and giving expert, constructive advice, and the English Heritage Trust is a charity caring for the National Heritage Collection of more than 400 historic properties and their collections.

First published 2000 (1-873592-57-4)
Second edition 2009 (978-1-84802-038-2)
Third edition published 2015

ISBN 978-1-84802-296-6

British Library Cataloguing in Publication data
A CIP catalogue record for this book is available from the British Library.

Historic England holds an unparalleled archive of 12 million photographs, drawings, reports and publications on England's places. It is one of the largest archives in the UK, the biggest dedicated to the historic environment, and a priceless resource for anyone interested in England's buildings, archaeology, landscape and social history. Viewed collectively, its photographic collections document the changing face of England from the 1850s to the present day. It is a treasure trove that helps us understand and interpret the past, informs the present and assists with future management and appreciation of the historic environment.

For more information about images from the Archive, contact Archives Services Team, Historic England, The Engine House, Fire Fly Avenue, Swindon SN2 2EH; telephone (01793) 414600.

Brought to publication by Robin Taylor, Publishing, Historic England.
Typeset in Source Sans Pro 9.5pt/13.5pt
Edited by Sparks Publishing Services
Page layout by Hybert Design, UK
Printed in the UK by 4edge Ltd.

Historic England

Contents of Metric Survey Specifications for Cultural Heritage

Acknowledgements

The preparation of this specification would not have been possible without the work of those, past and present, charged with the survey of the historic estate in the care of the Historic Buildings and Monuments Commission. This edition has been substantially revised and re-written but is based on the original document written by Jon Bedford, Bill Blake and Paul Bryan, of the then English Heritage Metric Survey Team with contributions from Dr David Barber and Dr Jon Mills (School of Civil Engineering and Geosciences, Newcastle University). Grateful thanks are also due to former members of the Metric Survey Team – Mick Clowes, Andy Crispe, Sarah Lunnon, Heather Papworth and Steve Tovey.

Preface

The survey brief

The preparation of a brief for the supply of survey services based on the options in this specification should ensure the necessary communication between the information user (the client) and the information supplier (the surveyor) required for the successful application of metric survey.

Performance of metric survey in heritage documentation

In order to obtain metric survey fit for the purposes of heritage management it will be necessary to consider not only the metric performance of measured data but also the required quality of work needed to act as both a record and an archive of the cultural heritage. The conventions of selection and presentation of measured drawing in architecture constitute a visual language that requires careful consideration. This specification contains both descriptions and illustrations of the required standard.

Use of the specification

This document is a description of the services and standards required for the supply of various types of metric survey. Sections 1, 2 and 3 describe the general terms, performance and presentation requirements common to all services. Sections 4, 5, 6, 7 and 8 contain standards specific to image-based survey, measured building survey, topographic survey, laser scanning and building information modelling (BIM), respectively. The use of any part of this specification without reference to the appropriate clauses of sections 1, 2 and 3 plus the appropriate service description from parts 4, 5, 6, 7 and 8 will be a misuse of the document and is very likely to result in an unsatisfactory product. While it is hoped that this specification will be distributed widely and is available for anyone to use, its use is not a guarantee of the required results and it is recommended that, if in doubt, professional advice is sought.

Structure of the document

This document comprises both a specification and guidance notes. The contents of the right-hand pages constitute the clauses of the specification, while the left-hand pages contain guidance pertinent to those clauses. Where the title of a clause on a left-hand page is prefixed by a # symbol, then the clause requires intervention from the user of the document. This usually takes the form of making choices from a list by deleting the options that are not required, but may also require ticking particular options in a list or the insertion of text. If a clause is not edited, then the first option should be taken as the default. In most cases this will be the recommended option.

Section 1

General conditions and project information

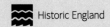
Historic England

1.1 Project brief

1.1.2 #Purpose of project

Give a brief description of the type of survey required and the intended end use.

1.1.3 #Location

Give as much detail as possible including a full address with postcode.

1.1.4 #Access arrangements

Give opening hours, where applicable the name and telephone number of a site contact, plus anyone else who would need to be informed of, or give permission for, attendance on site.

1.1.5 #Health and safety statement

Make reference to the client or site owner's health and safety policy, any existing risk assessment or any particular hazards the contractor should be aware of.

1.1.6 #Copyright

State who is to retain copyright of all materials resulting from the survey.

1.1.7 #Contract

State the conditions of contract that are to apply to the project. If the client does not have their own conditions it is recommended that the following is used:

> RICS, 2009 *Terms and Conditions of Contract for Land Surveying Services*, 3 edn. London: RICS, available from the RICS Shop: www.rics.org/uk/shop/

1.1.8 #Site clearance

State whether or not the site is free from obstructions and whether any clearance will be undertaken prior to the survey commencing. This is particularly relevant for image-based surveys.

1.1.9 #Completeness of survey

State, given the above, the degree of completion required and whether field completion by other means will be required for image-based surveys.

1.1.10 #Area and scale of survey

Describe the area, type and scale of survey required as comprehensively as possible, preferably with the aid of a map.

1.1.11 #Delivery schedule

Give the final deadline plus any phasing of delivery that may be required. Allowance will need to be made for lead times, fieldwork and office processing.

1.1 Project brief

1.1.1 Name of project and reference number

1.1.2 Purpose of project

1.1.3 Location

Address ...

Directions ...

National Grid Reference ...

1.1.4 Access arrangements

1.1.5 Health and safety statement

1.1.6 Copyright

1.1.7 Contract

The conditions of contract are ...

1.1.8 Site clearance

1.1.9 Completeness of survey

1.1.10 Area and scale of survey

1.1.11 Delivery schedule

Historic England

1.2 Introduction

Please note that guidance with regard to legislation and regulations is specific to England and will vary in different countries.

1.2 Introduction

The general conditions cover aspects of undertaking survey that are common to most metric survey activities carried out on historic sites. The project brief consists of the administrative and logistical aspects of a particular project.

1.2.1 Pertinent legislation

Contractors are to be aware of all current statutory requirements relevant to the contract for survey work. The contractor's attention is brought to:

- the Ancient Monuments and Archaeological Areas Act 1979;

- the Planning (Listed Buildings and Conservation Areas) Act 1990;

- the Construction (Design and Management) Regulations of 2015, introduced under the Health and Safety at Work Act 1974.

Copies of the above legislation can be obtained from The Stationery Office:

www.tsoshop.co.uk

tel: 0870 242 2345

Where the survey work occurs in countries other than England, the law of that country will apply.

1.2.2 Client's guidance on matters concerning survey

Contractors are required to comply with the client's guidance on matters of safety and standards of work regarding the historic fabric.

1.3.3 #Risk assessment

A risk assessment should be requested with all quotations or tenders and will be essential for all projects involving small unmanned aircraft (SUA).

The contractor should be made aware of any known hazards before they attend site.

1.3 The contract and other documentation

1.3.1 Contract

The contract will consist of the conditions noted in section 1.1.7 of the project brief, this specification including any edits made specifically for the project plus any attached documents or diagrams.

1.3.2 Method statement

In response to a request for a quotation or invitation to tender, a method and resource statement is to be provided by the contractor. As a minimum it must include:

- method proposed for providing survey control and the required detail;

- number of and positions of staff to be employed on project, including project leader;

- survey equipment, cameras etc to be used;

- access equipment to be used;

- lighting and electrical equipment to be used;

- any proposed alternative survey methods and their performance;

- the anticipated level of possible 3-D site completion;

- proposed output device, resolution and media;

- data retention and archiving arrangements; and

- delivery schedule.

1.3.3 Risk assessment

A risk assessment must also be supplied with the quotation or tender.

Known hazards include but are not limited to............(specify).

1.3.4 Site visits

The contractor may wish to visit the site to verify the requirements of the project and facilitate the production of the quotation or tender, the method statement and the risk assessment.

Where access to land not in the client's care is necessary, assistance will be provided to secure the appropriate way-leaves.

1.3.5 Calibration certificates

Copies of up-to-date calibration certificates for all relevant equipment are to be supplied with the quotation or tender.

1.4 Contractual details

1.4.1 Completion of survey

The client will seek agreement with the contractor on the extent of cover, within the acceptable limits of tolerance and method (ie establishing any areas that require an alternative survey technique or that cannot be covered). Where obstructions to survey exist, the client will seek agreement about the possible extent of completion.

1.4.2 Right of rejection

The client reserves the right to reject the application of any proposed survey technique or submitted survey product.

1.5.2 #Health and safety requirements

Edit as appropriate.

1.5 Health and safety

1.5.1 Contractor's responsibilities for safety

The following requirements are included here as a guide, and contractors must ensure that all relevant safety requirements associated with the provision of survey on behalf of the client are met during the contract period. The contractor's attention is brought to the need for best practice in matters of safety.

1.5.2 Health and safety requirements

The client's health and safety requirements are either:

(a) found at 1.1.5; or

(b) attached to this document.

1.5.3 Health and Safety at Work Act 1974

Under this Act employers have responsibilities to their employees and those affected by their work (eg members of the public and staff on the site). Further information on this can be obtained from:

> Health and Safety Executive:
>
> www.hse.gov.uk

Publications may be ordered from HSE Books:

> books.hse.gov.uk/hse/public/home.jsf
>
> tel: 01787 881165

1.5.4 Access equipment

Access equipment supplied or used by contractors or their agents must conform to the current safety standards. The contractor's attention is drawn to:

- Work at Height Regulations 2005 SI 2005/735 The Stationery Office 2005, ISBN 0 7176 2976 7; www.legislation.gov.uk/uksi/2005/735/contents/made

- Lifting Operations and Lifting Equipment Regulations 1998. Found in Safe Use of Lifting Equipment, available from HSE Books, ISBN 0 7176 1628 2.

- Provision and Use of Work Equipment Regulations 1998. Found in Safe Use of Work Equipment, available from HSE Books, ISBN 0 7176 1626 6.

Full details, certification and nominated safety contacts on proposed access equipment, where relevant, are to be included in the method statement. The contractor is to comply with English Heritage *Management Guidance:* Safe Use and Maintenance of Ladders (*see* Appendix 1.3).

1.5.5 Electrical equipment

Where applicable, electrical equipment (the use of domestic/battery-operated equipment is not included) must meet the requirements of the Electricity at Work Regulations 1989. Found in the Memorandum of Guidance on the Electricity at Work Regulations 1989, available from HSE Books, ISBN 0 7176 1602 9.

The contractor will comply with English Heritage management guidance: Portable appliance testing (*see* Appendix 1.2).

1.5.6 Survey equipment

Survey instruments or associated laser pointing devices, which may be a health hazard to people working in or visiting the site during the project, must be included in the risk assessment. Any certificates or statements from the manufacturers concerning the safety of the equipment must be included in the assessment along with any requirement for notification of 'lasers in operation' on site. English Heritage requires users of all such equipment to comply with BS EN 60825-1: 2014 *Safety of Laser Products Part 1: Equipment classification and requirements.*

1.6.1 Use of ground markers

For further details on obtaining scheduled monument consent in England, *see* the following link to the Historic England website:

www.historicengland.org.uk/advice/hpg/consent/smc/

1.6.2 Surface-mounted targets

For example, A4-size checkerboard targets should not be fixed to historic fabric with high-tack adhesive tape. Indeed, for laser scanning it will be preferable to use targets or reference objects mounted on stands or tripods. These will avoid the possibility of damage to historic fabric and can be positioned so as not to obscure detail.

1.6 Damage to site and fabric

Contractors are reminded that there is a range of penalties and powers of prosecution under the provisions of the Ancient Monuments and Archaeological Areas Act 1979 and the Planning (Listed Buildings and Conservation Areas) Act 1990 should unauthorised work be carried out or damage be caused to the building or monument.

1.6.1 Use of ground markers

The use of nails, permanent station markers, etc is subject to approval of the mark and its location. The insertion of any mark may require scheduled monument consent (SMC) and must not be done without the permission of the client.

1.6.2 Use of surface-mounted targets

Surface-mounted targets, such as for photographic or laser scanning control, must be no larger than 200mm by 200mm and must only be fixed with an approved adhesive that will allow removal without damage to the surface. *See* sections 4.2.2 and 7.2.11 for further details.

1.7 Survey material supplied

1.7.1 Copyright

The copyright of all materials generated as part of the contract is to be transferred to the client unless stated otherwise in section 1.1.6.

1.7.2 Retention of survey documentation

On request the contractor shall make available to the client all materials used for the compilation of the required survey. This material must be retained by the contractor for a minimum of seven years.

As a minimum this material will include: field notes and/or diagrams generated while on site; the raw and processed data used for the final computation of control; and a working digital copy of the metric survey data that forms each survey drawing or model (including formatted 2-D and 'raw' 3-D data files). The precise digital format and file type of this archive will be specified in section 3.1. If during this period the contractor wishes to change the format of this data archive, they are to seek the client's permission.

Appendix 1.1 Hot works: management guidance

General

As a general rule *no* hot work will be permitted in any roofed historic building in the care of English Heritage. This includes all operations involving flame, hot air, arc welding and cutting equipment, blowlamps, bitumen boilers and other equipment producing heat or having naked flames.

However, in exceptional circumstances when there is absolutely no alternative, the Conservation Maintenance Manager may allow hot work to be carried out on strict adherence to the provisions of the Hot Work Permit. Increased costs for carrying out the work without the use of 'hot work' must not be a deciding factor in the granting of a Hot Work Permit.

Many serious fires in historic buildings have been caused by contractors using heat-producing equipment. These fires are avoidable and our policy is to avoid hot works altogether unless there is absolutely no alternative.

Requirement 1

The Hot Work Permit must contain a risk assessment, a method statement, authority to start the works, an audit provision and a procedure for checking the area after completion of works.

The standard hot works permit form, properly filled in, will enable requirement 1 to be met. The individual items on the form should only be completed at the specified time.

Requirement 2

The Responsible Person must not allow hot works or naked flames to take place without a current hot works permit. The works must be finished early enough for the area to be checked two hours after completion.

The Responsible Person as identified in Management Standard MS001 must be firm about not allowing hot works in their property without a completed Hot Work Permit. They are also responsible for checking that the conditions of the permit are being met.

If, however, a project involving, for instance, a new visitor centre which is separate from the historic building is being undertaken, it will be the person managing the project that will be responsible for ensuring that the Hot Work Permit system is being adhered to.

Requirement 3

The use of gas cylinders of any type is prohibited within Historic Properties unless the cylinders are in a locked cage external to the property in a safe location and the gas supply is piped into the building by an English Heritage approved system.

Liquefied petroleum gas (LPG) is commonly sold in cylinders as butane or propane. It is heavier than air and is explosive when mixed with air in relatively low concentrations. This poses a serious danger to people in and around the property, particularly in places such as basements where leaking gas could accumulate. For this reason gas cylinders should be stored externally where leaking gas can dissipate safely. The use of LPG heaters should be avoided for fire safety reasons and also to avoid the large amounts of condensation produced.

Requirement 4

Where hearth fires are lit within a roofed historic building, a Hot Work Permit, approved by the Conservation Maintenance Manager, may cover an annual programme of use, provided that regular inspections and checks of the hearth and flue by the Regional Facilities Team form part of the method statement.

Appendix 1.2 Portable appliance testing: management guidance

Introduction

Portable appliance testing is commonly referred to as PAT testing (even though the word 'test' is represented twice). It refers to the principal means of ensuring the safety of portable electrical appliances.

There is no direct legal requirement for PAT testing. However, the Electricity at Work Regulations require electrical appliances to be maintained to prevent danger. Therefore PAT tests are carried out to determine if an appliance needs maintenance in order to ensure safety.

Requirement 1: Safety of appliances

All electrical appliances under the direct control of English Heritage must be maintained to prevent danger to the user or the building in which the appliance is used. The need for maintenance must be identified by carrying out periodic testing (in-service test) by a competent person. The test must include:

- a preliminary inspection;

- earth continuity tests (where required);

- insulating testing/earth leakage measurement; and

- functional checks.

The phrase 'under direct control' is used because English Heritage is responsible for the safety of all employees who use portable appliances, regardless of who owns the appliance. Therefore English Heritage must arrange for PAT tests on hired equipment used by English Heritage employees, where the equipment is under a long-term lease agreement. Where the equipment is leased in the short term, or borrowed from a third party, English Heritage must ensure that the appliance has been maintained by the owner or supplier before it is used by an English Heritage employee. Therefore if an appliance is used by an English Heritage employee, at the point of use it is under the direct control of English Heritage and must be safe.

In order to fulfill this requirement, the following questions need to be addressed:

- What needs to be tested?

- Who is responsible for arranging the test and making sure appliances are maintained to prevent danger?

- Who should carry out the test?

- How often should the test be carried out?

Each of these will be considered in turn.

What needs to be tested?

The following need to be the subject of a PAT test:

- portable appliances, eg mobile phone chargers, CD players, projectors;

- movable equipment, eg toasters, desk lamps, kettles;

- handheld equipment, eg cordless drills, hair dryers, powered screwdrivers;

- stationary equipment, eg photocopiers, microwave ovens, fridges; and

- IT equipment, eg docking stations, desktop PCs, monitors etc.

It includes English Heritage catering equipment and equipment used by gardeners. Usually the person undertaking the test will decide what should be tested. It is usual practice for the tester to progress through a workplace testing all appliances that they find. A rough rule of thumb is that an appliance is any piece of electrical equipment that draws power from a main circuit via a plug and socket. Therefore at the time of the test, the tester needs to have access to all appliances, otherwise the test will be incomplete and employees who use untested appliances could be at risk. It is best practice to maintain a register of appliances. This is simply a list of appliances. New appliances should be added to the register as they are acquired. Usually the register is held and maintained by the person who arranges the PAT testing. Where a register is used, everyone within the building must contribute to maintaining its accuracy by informing the keeper of acquisitions and disposal.

Who is responsible for arranging the test and making sure appliances are maintained to prevent danger?

As with all health and safety issues, responsibility runs through the line management structure. Line managers are responsible for ensuring that the work carried out by their employees is reasonably safe. This is true for the safety of electrical appliances. However, in practice the line manager rarely arranges for the test or maintenance to take place. Within offices, the Administrative Estates Department or Office Manager (or similar) usually arranges for the appliances within the office (as well as the fixed electrical systems) to be tested. Within visitor sites, the Estates Department (via the M&E Technical Manager) usually arranges for the appliances (and fixed systems) to be tested. The appliances used by homeworkers must also be tested. Again, line managers are responsible for ensuring that the appliances are safe. This responsibility can be fulfilled by making arrangements either to take the appliances to the nearest English Heritage office when a PAT test is due at that office or for a PAT tester to visit the homeworker at home. If appliances are to be tested with an office then the arrangements need to be made with the Administrative Estates Department or Office Manager. If the test is to take place within the home of the homeworker, then arrangements either need to be made directly with a suitable tester or it may be possible to make arrangements to use the tester employed by the Estates Department to undertake tests at visitor sites. If English Heritage employees work at third-party sites then the line manager should make enquiries with the person or organisation who controls the site to make sure that any appliances to be used by the English Heritage employee have been maintained to prevent danger.

Who should carry out the test?

The test must be carried out by a competent person. This means that the person must have the appropriate knowledge and experience to be able to carry out the test and recognise faults or other indicators of potential danger. There are no specific training requirements to undertake a PAT test, but the tester should be familiar with the current edition of BS 7671, the Electricity at Work Regulations 1989 and the Institution of Electrical Engineers (IEE) code of practice for In-Service Inspection and Testing of Electrical Equipment. In practice they must have undertaken appropriate training and have experience of carrying out PAT tests, including;

- a preliminary inspection;

- earth continuity tests (where required);

- insulating testing/earth leakage measurement; and

- functional checks.

How often should the test be carried out?

There are no legal standards relating to the frequency of testing. Some electrical contractors will claim that an annual test is required and will apply stickers to appliances warning users not to use it after 12 months. This is misleading and could cause unnecessary alarm. The Health and Safety Executive recommend the following intervals:

- *information technology*, eg desktop computers, VDU screens etc: formal visual inspection every 2–4 years and a test up to 5 years;

- *photocopiers, fax machines*, not handheld, rarely moved: formal visual inspection every 2–4 years and a test up to 5 years;

- *double-insulated equipment*, not handheld, moved occasionally, eg fans, table lamps etc: formal visual inspection every 2–4 years;

- *earthed equipment (Class 1)*, eg electric kettles, some floor cleaners: formal inspection every 6 months to a year, with a test every 1–2 years;

- *cables (leads) and plugs connected to the above* or extension leads: inspection every 6 months to 4 years depending on use and environment, a test is required every 1–5 years depending on the same.

These are only suggestions. More frequent inspections and tests will result in an increased level of protection. However, it is not necessary to use resources to test low-risk items on a frequent basis.

Requirement 2: Making appliances available

Where arrangements are made to carry out portable appliance testing, employees must, with reasonable notice, ensure that the electrical appliances they use are available for testing. In order to ensure that an arranged test covers as many appliances as possible (subject to whether or not they need to be tested, *see* above), it may be necessary to ensure that the tester has access to equipment in cupboards or other storage areas. In most cases, the equipment will be locked away, therefore employees who control access to the equipment must ensure that the tester can access and carry out a test. This requirement refers to 'reasonable notice'; the person who arranges the test must give reasonable notice of the date of the test and allow employees to take appropriate action to allow access. Consideration must be given to maintaining security and preventing theft. In some circumstances, the tester may need to be supervised when accessing equipment, for example because the equipment involves hazardous substances or is located in a sensitive area.

Requirement 3: User inspections and reporting problems

Any English Heritage employee who uses electrical appliances must carry out informal visual inspections to identify obvious defects. If defects are identified then they must be reported and appropriate action taken.

Anyone who uses electrical appliances should carry out simple visual inspections of the appliance. The inspections should take place on a routine basis. In practice, this happens all the time. Users tend to notice obvious and significant defects such as:

- scorch marks (and the smell of burning);

- frayed cables;

- cracked plastic casing; or

- loose or exposed cables (including seeing sparks or arching).

There is no set procedure for carrying out user inspections; each one will depend on the nature of the appliance. In addition to this, no specific competence is required beyond a basic awareness of common or typical defects (such as the above examples). If a defect is identified then the appliance must be taken out of use if the defect poses a danger (or has the potential to pose a danger). Usually the person who identifies the defect is responsible for taking the appliance out of use. This can easily be achieved by either removing the power lead (eg from a kettle) or by simply attaching a notice to the appliance informing people of the defect and warning them not use it. The person who discovers the defect should report the nature of the defect to the person who can arrange for maintenance (for example, the M&E Technical Manager in Estates, a Facilities Manager, an Office Manager, the supplier of leased equipment). The request should be made in writing, taking account of local protocols (for example works requests). Any defects identified during formal inspections and testing by a competent person, in response to Requirement 1, will be controlled as part of the inspection process arranged with the competent person.

Requirement 4: Maintenance of records

A record of all tested appliances must be provided by the competent person who carried out the test, including details of:

- the identification of the appliance;

- its status in terms of the test requirements; and

- recommendations where an appliance fails and requires maintenance to prevent danger.

A copy of the record must be maintained in the workplace. Where appropriate, the appliance can be fixed with a sticker to indicate that it has been tested.

In practice, most PAT testers produce a computer printout detailing all the appliances they have tested. This should form an inventory or register of appliances within the workplace. Upon receipt, it should be maintained by the person who arranged the PAT test programme, using information supplied by the users relating to the acquisition and disposal of appliances.

The inventory should be kept in the workplace in order to form a record of the test and so that if an incident occurs where an appliance poses a danger, the inventory can be examined to determine if and when the appliance was tested.

Further information on the safety of electrical appliances and PAT tests can be obtained from the following:

- the Territory Health and Safety Coordinators;

- the Building Services Engineering and Safety Team; or

- the M&E Technical Managers in the Estates Department.

Appendix 1.3 Safe use and maintenance of ladders: management guidance

Legal requirements

The Working at Height Regulations 2005 include the use of ladders as access or a working position.

The Provision and Use of Work Equipment Regulations 1998 require that ladders, along with all other work equipment, are suitable for the use that will be made of them, and are efficiently maintained and repaired.

The following requirements should be met when using a ladder:

- Any ladder shall be suitable and of sufficient strength for the purposes for which it is to be used.

- A ladder should be so erected as to ensure that it does not become displaced.

- If it is more than 3m long, a ladder should be secured if practicable. If not practicable, a person must be positioned at the foot of the ladder to prevent it slipping at all times when it is being used.

- The top of any ladder used for access to another level must extend to provide adequate handhold.

- Ladder runs in excess of 9m must be provided at suitable intervals with landing areas or rest platforms.

- Any surface upon which a ladder rests shall be stable, level and firm, of sufficient strength and of suitable composition safely to support the ladder and any load intended to be placed upon it.

Whilst these requirements are for 'construction' work, they provide a good guide to the safe use of ladders in other work scenarios.

Safe use of ladders

Except for very short duration tasks ladders should not be used as a 'place of work', and only then if the user can carry out the task without over-reaching. They should always be secured, but if this is impracticable, they must be footed by a second person. Reaching sideways from a ladder must be avoided as there are severe hazards of the ladder slipping. Standing on a ladder for lengthy periods is also very uncomfortable on the feet. Care should be taken when using an extension ladder, particularly when descending past the overlap, as it is very easy to miss your footing.

Portable step ladders

These should also only be used for very short-term access, rather than relying on them as a work platform. Always ensure that the ladder is fully opened and that the restraining cord is in good condition. The surface underneath must be flat and firm. Never climb beyond the point where you have a secure handhold. For regular use in stores etc, step-ladders with built in hand rails should be used. (See below for safety advice when using ladders and step ladders.)

Maintenance procedures

To meet the statutory requirement in respect of such items of plant used by English Heritage it is the responsibility of Maintenance Managers to ensure that:

- all items are regularly inspected by a competent person; and

- records of these inspections together with details of the condition are kept up to date in a register, (*see* below).

Competent person

To be considered a 'competent person' that person should possess the technical knowledge, practical experience and ability to carry out this function. It is not essential that this person is a tradesman.

Identification

Each timber or metal step ladder, trestle, crawling ladder, crawling board and component tower must be marked with a serial number for identification purposes, together with the date at which it is next due for inspection.

Inspection

Each item is to be inspected upon receipt from the supplier, and at 6-monthly intervals. This period may be reduced in the light of local conditions and frequency of usage at the discretion of the local manager.

Inspection should include checks for:

- corrosion, warping, cracking and splintering of rungs;

- wear and condition of rungs, especially at stile fixing;

- tightness of wedged joints;

- loose corroded tie rods and reinforcing;

- corrosion, wear and security of hinges, check stops, locking stays, cords, screws and bolts;

- condition of operating cords;

- corrosion, wear and security of pulley wheels, latch locks and guides;

- correct fitting and wear of pads on levelling and non-slip feet attachments;

- condition of varnish or other protective coating (paint must not be used); and

- legibility of identification mark.

Keeping of records

A register is to be maintained in the manager's office and should include the following information, recorded on the appropriate planned maintenance log book sheet;

- description

- identity mark;

- date of receipt into store;

- date of periodic examination;

- name and grade of person carrying out inspection; and

- condition at time of inspection, ie serviceable or details of defects found.

Defective equipment

Any item found to be defective is to be labelled as such and impounded until remedial or disposal action can be taken. Items beyond economical repair must be destroyed. Details of remedial action and disposal are to be entered in the register.

Safety notes for working on step ladders

For example as access or as a place of work for painting, decorating, ceiling work, electrical wiring, light fittings etc.

Hazards

- *Over-reaching:* The steps can become unstable and you, the steps and any tools can fall and possibly strike persons below.

- *Sideways loading eg drilling:* You push yourself off the steps and fall, and you, the steps and tools possibly strike persons below.

- *Losing balance:* You grab the steps and they become unstable; you, the steps and any tools fall, possibly striking persons below.

Check:

- Is the step ladder capable of reaching the working height?

- Is there any wear, tear or damage affecting safety (eg hinges, rivets and dents)?

- Can you carry the step ladder and position it safely without slipping, gripping or falling?

- Is your footwear OK for working on a ladder?

- Are ground conditions firm level and stable?

- Do weather conditions allow safe use of the step ladder (eg wind, rain snow, ice, lightning)?

- Duration of task without a break not to exceed 30 minutes?

- Can you carry any materials, equipment or tools safely while maintaining a handhold?

- Can you place steps to avoid over-reaching and sideways-on loading for each task?

- Could sudden unexpected movement cause you to fall off the step ladder, and how serious would be the consequences?

- Should you segregate the work area to protect any other personnel?

Appendix 1.4 Scaffolding: management guidance

Introduction

Scaffolding may be erected in or around an historic building or monument for various reasons. These may include access for inspection of the structure, repairs to the structure, cleaning, painting, recording etc and also for shoring purposes and temporary works.

The problem with scaffolding historic buildings and monuments is that they are seldom regular shapes like new buildings, and the structure being scaffolded may be in an unstable condition, requiring great caution during erection and possibly ruling out the option of tying the scaffold to the structure. Badly erected or misused scaffolding can lead to serious or even fatal consequences, even with quite minor structures.

Scope

This Safety Instruction covers all types of scaffolding and temporary works erected on English Heritage sites. It outlines the requirements for the provision, construction, inspection and safe use of scaffolding and temporary works.

Provision

Scaffolding must be provided where there is a requirement to provide a safe place of work, if access cannot be safely gained by other means. The Construction (Design and Management) Regulations 2015 require safe access to be planned, and places specific duties on the Principal Contractor regarding the provision and maintenance of safe access on site. It is not acceptable to carry out work from ladders other than for the most minor tasks. Proprietary tower scaffolds may be used in some instances, but generally a full tube and coupling scaffolding should be provided.

Compliance with British Standards

All scaffold structures erected at English Heritage sites are to comply with the relevant British Standards, ie BS 1139: 1990 for Metal Scaffolding, BS 5973: 1993 Code of Practice for Access and Working Scaffolds in Steel and BS 5974: 1990 Code of Practice for Temporarily Installed Suspended Scaffolds and Access Equipment.

Design

All access scaffolding and temporary works should be designed by a competent person. In minor cases design can be by inspection of the job in hand, backed by experience. Larger schemes should be designed by a competent scaffolder or by a structural engineer.

Purpose

The use for which a scaffold may be put while it is standing must be considered. For example, although the initial intention may be to just inspect a structure, this may lead on to repair work being undertaken while the scaffolding is available. Sufficient load-bearing capacity should therefore be built into the scaffolding and platforms should be made to full working widths.

Materials

Although steel is the main material used in the scaffolding industry, aluminium alloy tubing, which imposes lower loads on archaeologically sensitive ground areas, may be specified for certain projects. It is important not to mix steel and alloy within a structure due to the differing load-bearing characteristics of the two metals.

Erection

Scaffolding must only be erected by specialist contractors whose employees must first have attended a CITB scaffolders' course, and have received a certificate of competence. The course is in two parts, and at least six months' work experience should be gained between taking parts I and II.

Details of proposed scaffolding should be approved by project officers well before work starts on site. The scale of the details will be commensurate with the scale of the proposals, and can range from a hand-drawn sketch in simple cases to a full set of detailed drawings in complex cases.

Inspections

All access and working scaffolds must be inspected weekly (or after inclement weather) to ensure their continued safety.

The inspection should be carried out by a competent scaffolder, or a supervisory grade with the necessary experience to carry out the inspection competently.

Records of the inspection are to be kept on site for inspection by any enforcement officer.

General safety requirements

Foundations

All access scaffolds and temporary structures are to be properly founded. Where it is intended that the construction is to remain in existence for any length of time, railway sleepers or similar-sized treated timbers should be used as sole plates in preference to scaffold boards. It is necessary to ensure that the ground or structure on which the scaffold or temporary structure sits is capable of carrying the load applied. Particular attention must be given to the possible existence of sub-surface drains or other voids.

Standards

Perhaps the most important point of concern regarding standards and other vertical members, apart from the fact that they are properly founded, is that they are vertical. Particular attention to the plumbness of vertical members is vital and spirit levels should be used for this purpose. Out of plumb standards are weaker than vertical standards and out of plumb vertical members induce horizontal forces on the structure as a whole.

Bracing

All scaffold structures require bracing to ensure that they maintain their structural integrity. Bracing will almost certainly be required in both directions and additionally plan bracing will sometimes be necessary. Bracing connections should be made close to standard/ledger/transom junctions. Where facade bracing is used, joints in tubes must only be made with sleeve couplers, never with joint pins.

Tying

Badly fixed, incorrectly positioned or inadequate numbers of ties are very frequent problems with scaffolding. All scaffolding structures need to be tied to the permanent structure to ensure their structural stability. In exceptional circumstances buttresses can be built to provide stability to a scaffold. However, the compression members in buttresses must be braced in two directions, not just one as is often done.

Platforms

Care must be taken to ensure that decayed, warped or split boards are not used and also care must be taken to ensure that the positioning of the transoms does not create traps. It is very often necessary to ensure that boards are held down against wind forces.

Care must be taken to ensure that platforms are not overloaded by stored materials.

Guardrails

All working platforms at a height of 2m or more must be fitted with a guard rail at least 910mm above the platform, an intermediate rail and toe boards, such that the maximum opening is 470mm.

Protection

Where there is a risk of falling materials, brick guards should be provided and if necessary specially designed fans constructed.

Sheeting

Sheeting must not be fixed to any scaffolding without approval unless it has been specifically designed for that purpose.

Access

Safe access must be provided to all scaffolds, whether by ladder, ramp, link to building etc. All ladders must be secured, and generally inclined at a 4:1 angle. They should be removed or made inaccessible when the site is unattended.

Lightning protection

All scaffolding structures are to be bonded to earth by means of suitable earth electrodes in accordance with BS 6651, to provide lightning protection.

Stored materials

Scaffolding materials stored at depots and on sites are to be the subject of periodic inspection in accordance with normal plant maintenance procedures.

Appendix 1.5 Laser scanning health and safety considerations

The International Electrotechnical Commission standard *Safety of Laser Products – Part 1: Equipment classification and requirements* (IEC 60825-1: 2014) or the UK implementation (BS EN 60825-1: 2014) provide information on lasers and describe precautions in the use of laser products. Users should refer directly to this document when preparing health and safety assessments. However, a brief summary is provided below.

Dangers

Lasers used in survey applications may have risks associated with eye damage. The British Standard (BS EN 60825-1: 2014) defines seven relevant classes of laser (class 1C is only relevant to medical applications).

Class 1 lasers are safe under reasonably foreseeable conditions of operation, including the use of optical instruments for intrabeam viewing.

Class 1M lasers are safe under reasonably foreseeable conditions of operation, but may be hazardous if optics are employed within the beam.

Class 2 lasers normally evoke a blink reflex that protects the eye, this reaction is expected to provide adequate protection under reasonably foreseeable conditions, including the use of optical instruments for intrabeam viewing.

Class 2M lasers normally evoke a blink reflex that protects the eye, this reaction is expected to provide adequate protection under reasonably foreseeable conditions. However, viewing of the output may be more hazardous if the user employs optics within the beam.

Class 3R lasers are potentially hazardous where direct intrabeam viewing is involved, although the risk is lower than that for Class 3B lasers.

Class 3B lasers are normally hazardous when direct intrabeam exposure occurs, although viewing diffuse reflections is normally safe. This class of laser is generally not suited for survey applications.

Class 4 lasers will cause eye or skin damage if viewed directly. Lasers of this class are also capable of producing hazardous reflections. This class of laser is not suited for survey applications.

Users of laser scanning systems should always be aware of the class of their instrument. In particular the user should ensure the correct classification system is being used (ie IEC 60825-1: 2014 or BS EN 60825-1: 2014 rather than any previous version that differs slightly in classification).

Precautions

The British Standard (BS EN 60825-1: 2014) provides a number of safety precautions that should be observed during the use of laser scanning surveys. For lasers up to Class 3R (those normally used in survey applications) and where applicable to laser scanning for metric survey these precautions are briefly outlined below.

For a full description the user is referred directly to the British Standard, however, generally:

- Care should be taken to prevent the unintentional specular reflection of radiation.

- Open laser beam paths should be located above or below eye level where practical.

- Only persons who have received training to an appropriate level should be placed in control of laser systems. The training, which may be given by the manufacturer or supplier of the system, the laser safety officer or an approved external organisation, should include, but is not limited to: familiarisation with operating procedures; the proper use of hazard control procedures, warning signs etc; the need for personal protection; accident reporting procedures and bioeffects of the laser upon the eye and skin.

- Particular care should be taken with the use of magnifiers or telescopes around laser devices that may pose a risk when intrabeam viewing is used.

- The instrument should only be used in accordance with the manufacturer's instructions.

For lasers that emit energy outside the wavelength range of 400–700nm special considerations are often required. For example:

- Where using a Class 3R laser a laser safety officer should be appointed.

- Beam paths should be as short as possible and avoid crossing walkways and access routes.

Particular precautions and procedures are outlined in the standard for Class 1M, Class 2M and Class 3R laser products used in surveying, alignment and levelling. Those with relevance to laser scanning are:

- Only qualified and trained persons should be assigned to install, adjust and operate the laser equipment.

- Areas where these lasers are used should be posted with an appropriate laser warning sign.

- Precautions should be taken to ensure that persons do not look into the beam (prolonged intrabeam viewing can be hazardous). Direct viewing of the beam through optical instruments (theodolites, etc) may also be hazardous.

- Precautions should be taken to ensure that the laser beam is not unintentionally directed at mirror-like (specular) surfaces.

- When not in use the laser should be stored in a location where unauthorised personnel cannot gain access.

Other considerations

In addition to the risks associated with lasers, users should be aware that due to the size and weight of some systems there is a risk of injury to visitors, especially children, if systems are left unaccompanied.

The effect of laser scanning on features such as lichens and delicate fabrics is not well understood. Consideration should be given to the use of lasers in the vicinity of such features.

Section 2

General performance and control of metric survey

2.1 General performance requirements

This document is a performance-based specification that describes two types of survey product. These are:

- vector presentation in computer-aided design (CAD) software (*see* sections 4, 5 and 6); and

- image-based presentation (*see* sections 4 and 7).

It is intended for the generation of accurately located base survey data, including building information modelling (BIM) ready models, into which specific thematic input can be added if required.

The performance of the survey types is compared against the following criteria:

- measurement (accuracy and precision);

- selection of features; and

- presentation of the survey data.

2.1.1 Measurement performance

Accuracy and precision

The terms 'accuracy' and 'precision' are often used colloquially to mean the same thing but their definitions are different, and this difference is relevant to the surveyor. Accuracy is how well a measurement conforms to its true value. Precision is how repeatable a measurement is. So a survey instrument can be accurate (in that it returns a value close to the correct value for a measured point) but imprecise (because it returns different values each time a measurement is taken) or it can be precise (returning similar values with each measurement) but inaccurate (because the values returned are not close to the real value). Ideally, a survey instrument will be both accurate and precise, returning results that are close to the true value of the measurement that can be repeated with very similar values as long as conditions do not change significantly. It is important to note that achieving high precision does not necessarily mean high accuracy is achieved, because of the chance that bias has been introduced. For example, this can occur when a poorly calibrated instrument is used that may return consistent but incorrect measurements. For this reason, regular calibration and testing of a survey instrument is essential, to ensure that it is both accurate and precise (to within its measurement tolerances).

In surveying, further refinements to these concepts are also made. For example, 'absolute accuracy' refers to the accuracy of a measurement with regard to a particular coordinate system and 'relative accuracy' refers to how well measured points are placed relative to one another.

For further information on accuracy and precision in surveying, *see* the RICS Client Guide *Reassuringly Accurate*:

www.rics.org/Global/Reassuringly%20Accurate.pdf

2.1 General performance requirements

Metric survey techniques are required to deliver data that can be verifiably repeated. There are three aspects to the required performance of metric survey data. These are:

- measurement performance;

- feature selection performance; and

- presentation performance.

This specification is intended for the generation of base survey data, located accurately in its true 3-D position, to which specific thematic input or attributes can be added if required.

2.1.1 Measurement performance

Measurement performance may be considered in terms of both accuracy and precision.

Definition of accuracy

Accuracy describes the closeness between measurements and their true values. The closer a measurement is to its true value, the more accurate it is.

Definition of precision

In surveying, precision is taken to describe the consistency with which a measurement or set of measurements can be repeated.

Repeatability of capture method

Data capture must be by a method that can be repeated, to the appropriate order of precision, by the use of similar equipment and suitably qualified personnel. Therefore the proposed method must be fully and clearly described in the method statement.

2.1.2 Scale tolerance and point density

Scale tolerance can be specified in terms of standard deviation (sigma) or root mean square error (rmse). Standard deviation is based on the assumption that random errors will have a normal distribution around the mean. One standard deviation means there is about a 68 per cent probability that all measurements will have errors not greater than the mean value. For more information on this subject *see*

> RICS (2014) *Measured Surveys of Land, Buildings and Utilities* (RICS Guidance Note), 3 edn, ISBN 978 1 78321 064 0, available from the RICS Shop: **www.rics.org/uk/shop/**

Measurement techniques can be characterised by whether they are active or passive. With an active technique the required points are selected and measured in the field, for example, using a total station theodolite. Conversely, a passive technique involves the capture of a mass of data from which, if required, points are selected as part of a later process. Examples of this include the production of drawings from photogrammetry or laser scanning. Alternatively all the points, or a sub-sample, can be presented in the form of an image such as a rectified photograph.

Point sampling with active techniques or as a post-process with passive techniques can be varied because of the selective nature of the method. Point densities will be increased in areas of greater detail and decreased if the surveyor judges there to be less information to be recorded. Stating a maximum distance between points will allow the data to be used for digital terrain model (DTM) generation if required.

Mass capture techniques need to provide a point density that is at most half the size of the smallest object to be recorded.

Measurements derived from reflectorless electromagnetic distance measurement (REDM) can be incorrect if the beam is interrupted or affected by multiple return signals, such as when aimed at the edge of a wall. Monitoring the data to ensure that lines clearly describe the true edges of the building or monument is essential.

Historic England

2.1.2 Scale tolerance and point density

The precision of a survey is to be commensurate with the intended scale of presentation within the tolerances tabulated below. It is expected that surveyed data will allow repetition of a given measurement as presented on a plotted drawing within the following maximum tolerances when checked from the nearest control point.

Precision

Required maximum tolerance for precision of detail

scale	acceptable precision (1 sigma)
1:10	+/- 5mm
1:20	+/- 6mm
1:50	+/- 15mm
1:100	+/- 30mm
1:200	+/- 60mm
1:500	+/- 150mm

Point density/rate of capture

Required distribution of measured points

scale	point cloud	digitising*	field survey†
1:10	≤1mm	1–15mm (max 0.25m)	2–30mm (max 0.5m)
1:20	≤2.5mm	2.5–30mm (max 0.5m)	5–60mm (max 1m)
1:50	≤5mm	5–50mm (max 1m)	10–100mm (max 2m)
1:100	≤15mm	15–100mm (max 1.5m)	20–200mm (max 3m)
1:200	≤30mm	30–300mm (max 2.5m)	50–600mm (max 5m)
1:500	≤75mm	75–750mm (max 5m)	0.1–1.5m (max 10m)

* From photogrammetric model, laser scan point cloud or ortho-image.

† For example by total station theodolite (TST) or global navigation satellite system (GNSS).

In both cases where lines appear straight or detail is sparse the interval may be increased up to the maximum shown in brackets.

2.1.3 Completeness of survey

It is the responsibility of the surveyor to meet the requirements for coverage and completeness as set out in the project brief. The required extent of the survey should, however, be stated as clearly as possible by use of one or more of the following:

- site boundary marked on a sketch plan;

- elevations highlighted on a sketch plan;

- written description; and

- marked-up photographs.

2.1.3 Completeness of survey

The detail and precision with which survey data is collected must be commensurate with the required scale across the entirety of the survey, whatever the method or methods employed.

Survey coverage, with regard to both the extent of the survey and the completion required within that extent, is to be determined by the needs of the project. Elevations and sectional elevations shall be complete to full height unless otherwise specified. Any requirement for field completion of obscured areas by another method will be by agreement between the contractor and the client. *See* section 1.1.10 of the project brief.

2.2.1 #Accuracy of site control

Choose an option. Option (a) ±5mm will be suitable for most sites. For larger sites a part error may be more suitable. For example, 1 part in 20,000 for distances greater than 200m.

2.2.2 #Control network

Choose option (a) or insert own values at (b).

It will not always be possible to achieve this level of accuracy using GNSS alone. In such cases a TST traverse will be required. For the control of small unmanned aircraft (SUA) survey the accuracy of GNSS derived coordinate values will typically be sufficient.

2.2.3 #Existing coordinate system

Indicate whether there is a previously used site grid. If so, supply the necessary details as an attachment.

2.2 Control of survey

The control for all survey projects must be reliable, repeatable and capable of generating the required coordinates within the tolerances stated. The method, network and equipment for providing survey control are discretionary; however, details of the method and equipment proposed must be included in the method statement.

2.2.1 Accuracy of site control

The maximum error in plan between permanently marked survey stations after adjustment is to be no greater than either:

(a) ±5mm; or

(b) other (specify).

2.2.2 Control network

All coordinate and level values generated must be expressed in metres to three decimal places and presented in the order of easting (X), northing (Y) and height (Z). They are to be derived from a rigorously observed traverse and/or global navigation satellite system (GNSS) network to ensure that the following tolerances are satisfied, either:

(a) the horizontal closure error of any traverse shall not exceed ±10mm; the vertical closure error of any traverse shall not exceed ±20mm; or

(b) the horizontal closure error of any traverse shall not exceed......; the vertical closure error of any traverse shall not exceed......

Adjustments carried out to the observed network, including type and method of adjustment used and the results of transformations, are to be detailed in the final survey report.

2.2.3 Existing coordinate system

Where a previously defined site coordinate system exists, the necessary information will be supplied by the client to enable the re-occupation of permanently marked points. This will include a full listing of 3-D coordinates and witness diagrams. During re-occupation and re-observation the precision of any coordinate and level information provided must be evaluated to ensure the new survey can be generated within the appropriate tolerances. Where discrepancies are found, the client is to be contacted to agree any necessary variations.

There is either:

(a) an existing site coordinate system; or

(b) no existing site coordinate system.

2.2.4 #New coordinate system

Choose an option. If the data is to be used for geographical information system (GIS) applications or inserted into Ordnance Survey (OS) mapping then choose option (a) The scale factor, however, will distort facade control, so the use of a local grid derived from the Ordnance Survey National Grid (OSNG) coordinates will be required.

For more details on the use of GNSS and the OSNG, *see:*

www.ordnancesurvey.co.uk/business-and-government/help-and-support/ navigation-technology/os-net/surveying.html

Historic England, 2015 *Where on Earth Are We? The role of Global Navigation Satellite Systems (GNSS) in archaeological field survey*

available for free download from **www.historicengland.org.uk/images-books/ publications/where-on-earth-gnss-archaeological-field-survey/**

and for what to expect in a method statement refer to the following publication:

RICS, 2010 *Guidelines for the Use of GNSS in Land Surveying and Mapping*, 2 edn, ISBN 978 1 84219 607 6.

For guidance on the use of real-time kinematic (RTK) GNSS services, please *see* the following publications:

The Survey Association, 2013 *Guidance Notes for GNSS Network RTK Surveying in Great Britain*

Newcastle University (on behalf of The Survey Association), 2008 *An Examination of Commercial Network RTK GPS Services in Great Britain*

Newcastle University (on behalf of The Survey Association), 2012 *Further Testing of Commercial Network RTK GNSS Services in Great Britain*

available for free download from
www.tsa-uk.org.uk/for-clients/guidance-notes/.

2.2.5 #Vertical datum (height control)

Choose an option. The Ordnance Survey no longer maintains benchmarks and recommends that GNSS is used. For further details *see*
www.ordnancesurvey.co.uk/benchmarks/.

2.2.4 New coordinate system

If no previous survey coordinate system has been installed on site, either:

(a) the Ordnance Survey National Grid (OSNG) is to be used by means of GNSS observation. The WGS84 values are to be transformed to the OSNG using the OSTN02 transformation. Height values are to be transformed using the OSGM02 transformation. If the control is for an image-based, laser scanning or measured building survey, a local grid with no scale factor applied is to be derived from the OSNG values. Listings of both sets of coordinates are to be supplied; or

(b) a local coordinate system is to be established. The origin is to be positioned such that all grid values will be positive. The orientation is to be either as close to grid north – defined as the direction of a grid line that is parallel to the central meridian on the OSNG – as is practicable or parallel to the principal axis of the historic building or monument being surveyed.

2.2.5 Vertical datum (height control)

The vertical datum for the survey is to be:

(a) the OS height datum. This is to be achieved by means of GNSS observation and the OSGM02 transformation; or

(b) levelled to at least two local OS benchmarks. Where disagreements are found between benchmarks the client is to be contacted to agree any necessary variations prior to the survey continuing. The most recent OS height data to three decimal places and a location description of the bench marks must be included on the data sheet and/or the title box of each drawing sheet; or

(c) an arbitrary site bench mark. Full details of the site bench mark are to be included with the permanent survey mark witness diagrams.

2.2.6 #Establishment of permanent survey marks

Choose an option and if (a) edit as follows:

- insert the number required, if known;

- if not, insert 'at least 4' to allow for subsequent re-occupation; or

- delete the field and leave to the contractor's discretion.

If a local site grid is to be established for the first time it will be essential to have at least four permanent marks installed to allow subsequent re-occupation. Each marked survey station should be inter-visible from at least one other. Where an existing local site grid is to be re-occupied for work in a different part of the site, it will be prudent to have further permanent marks installed. If the OSNG is to be used it will still be useful to have permanently marked stations to allow re-occupation of the grid without the need for GNSS.

2.2.6 Establishment of permanent survey marks

Either:

(a) the establishment of [insert number] new permanent survey marks is required. Disturbance to the historic fabric must be kept to a minimum (*see* sections 1.6.1 above and 2.2.8 below); or

(b) the establishment of new permanent survey marks is not required.

2.2.7 Witnessing of stations

Full witness diagrams are to be provided with the survey for all permanently marked stations. Witness diagram sheets must include:

- coordinate values to three decimal places as eastings (X), northings (Y) and height (Z);

- a sketch diagram and dimensions to at least three points of hard detail;

- a written description of the mark; and

- a photograph of the location.

A traverse diagram must also be provided (*see* sections 3.3.6 and 3.4.5).

2.2.8 Use of ground marks

For further details on obtaining scheduled monument consent in England *see* the following link to the Historic England website:
www.historicengland.org.uk/advice/hpg/consent/smc/

2.2.9 Use of targets on historic fabric

If the fabric is fragile or historically significant, such as a wall painting or timber beam, guidance on suitable areas for target application should be noted here. A conservation professional should be consulted if there is any doubt about the suitability of a particular adhesive.

2.2.8 Use of ground marks

Permanent or temporary ground marks are to be as non-invasive as possible and preferably existing detail should be used. The type and location of any permanent mark must be approved by the client before insertion. In some cases scheduled monument consent (SMC) will be required.

The insertion of nails may require SMC. In any case nails must only be driven into a suitable material, for example earth, gravel or a mortar joint, not historic floorboards etc.

2.2.9 Use of targets on historic fabric

Where survey targets are to be applied to historic fabric, a suitable non-marking, non-destructive method of adhesion must be used. This must allow the removal of the targets without damage to, or marking of, the fabric. Details of the proposed method of adhesion are to be included in the method statement for the survey. The client reserves the right to refuse application if the proposed substance is deemed to be unsuitable for historic buildings or monuments. All targets must be removed before the commission is completed; any targets still remaining after completion will still have to be removed at the contractor's expense.

Section 3

Format, presentation and provision of survey data

3.1 Digital data

3.1.1 #CAD and digital image filenames

Choose an option or edit option (a) as required. For option (a) give the appropriate three-letter abbreviation.

3.1.2 #CAD data format

Choose an option. If image files are to be attached to the CAD files, ensure that the selected program and version supports this.

3.1.3 #Digital image format

Choose an option. TIFF files are recommended because they are not compressed and the format is non-proprietary. The Historic England digital archive only accepts images as TIFF files.

3.1 Digital data

3.1.1 CAD and digital image filenames

All CAD and digital image filenames are to be at least eight characters in length and must follow either:

(a) the following file naming system. The standard abbreviation for the site is ......

Characters	Description
1–3	Standard abbreviation of the monument name, eg DOV (Dover Castle)
4–5	Year survey/plotting carried out eg 15 (2015)
6	Type of survey
	P (photogrammetry – original images and 3-D CAD data)
	Q (photogrammetry – CAD drawing sheets)
	R (rectified photography – images and CAD files)
	O (orthophotography – images and CAD files)
	L (laser scan data)
	M (measured building survey)
	T (topographic survey)
7–8+	Sequential file number from 01 or 001 if there will be more than 99 files

eg DOV15T01.DWG, DOV15R01.TIF; or

(b) the system as described below (specify).

3.1.2 CAD data format

All CAD files, including any drawing sheets used to provide rectified photography or orthophotography are to be either:

(a) AutoCAD version....... .DWG; or

(b) other (specify).

3.1.3 Digital image format

Digital images are to be supplied either:

(a) as shown below

- the original digital images as 16-bit RAW files plus uncompressed 8-bit TIFF versions;

- where the images are rectified photography or orthophotography, as TIFF files; or

(b) as follows (specify).

3.2 CAD requirements

This specification is written for AutoCAD users. The following clauses may need to be edited for use with another CAD program.

3.2.1 #Use of CAD coordinate systems

Choose an option.

3.2.2 Insertion point

This is so that all drawing files for a survey relate to a common origin.

3.2.3 #CAD drawing unit

Choose an option. It is essential to know what units the coordinates represent so that dimensions are correct and plots are to the right scale. Architects and engineers, for example, may wish to have the drawing unit represent 1mm.

Assuming option (a), 1 drawing unit = 1m is chosen, then for each drawing units should be set as follows:

- Length – Type: decimal, with precision set to at least 3 decimal places;

- Angle – Type: decimal degrees with precision set to at least 3 decimal places; and

- Insertion scale set to either Metres or Unitless.

3.2 CAD requirements

3.2.1 Use of CAD coordinate systems

A user coordinate system (UCS) other than the world coordinate system (WCS) can be used to facilitate the presentation of the survey (or part thereof) on the desired sheet layout (*see* section 3.3.2 and figure 3.1 for further details). Any such UCS must be saved with a name related to its function (eg 'SHEETVIEW' for a UCS set-up for a drawing sheet).

Original 3-D data, without the addition of any title or border information, is to be provided set up in the following AutoCAD coordinate systems:

- the WCS related directly to the site (ground control) coordinate system;

- one or more UCSs, appropriately named, to enable the separate elevations to be viewed and edited directly as an orthogonal view. The origin of the UCSs must be such that all elements within the subject area are positive with low coordinate values (eg the lower left-hand control point given arbitrary values of 10m for the X and Z axes). The Y axis must be set to the true heights of the ground control.

As well as the original 3-D data, the data for each formatted drawing sheet is to be provided in either:

    (a) 2-D form only; or

    (b) 3-D form only with a suitable UCS.

Where possible all height datum lines or level ticks in a drawing are to have the correct height value on the vertical axis. Where two or more elevations are presented on the same drawing sheet, one above the other, then the datum lines or level ticks of the same value are to be separated by a whole number of metres.

3.2.2 Insertion point

The default origin of (0,0,0) in the WCS is to be used for xref insertions.

3.2.3 CAD drawing unit

The CAD drawing unit is to be either:

    (a) 1m; or

    (b) other (specify).

3.2.4 Line type

Where possible, a dashed line type is to be used for dashed lines as opposed to using a broken line. The line-type scale is to be commensurate with the plot scale so that it actually appears as a dashed line when plotted.

3.2.5 #Use of paper space

Choose an option. Unless it is set up as described in the specification, paper space can be confusing for non-specialist CAD users.

If paper space is not used, AutoCAD drawings are plotted from model space. In this case the following table of plot scale settings may be useful, assuming the drawing unit is 1m.

scale	plotted mm =	drawing unit
1:1	1000	1
1:5	200	1
1:10	100	1
1:20	50	1
1:50	20	1
1:100	10	1
1:200	5	1
1:500	2	1
1:1000	1	1
1:1250	1	1.25
1:2500	1	2.5
1:5000	1	5
1:10000	1	10

3.2.5 Use of paper space

Either:

(a) Paper space is to be used for the production of all drawing sheets and must be set up as follows:

- 1 plotted mm = 1mm in paper space.

- There is to be a specific paper space layout tab for each view or drawing sheet.

- The viewport(s) must be locked to prevent accidental changing of the scale.

- Where elevations are presented, one above the other, in the same layout separate viewports should be employed so that the elevations remain at their true height in model space (*see* section 3.2.1).

Or

(b) Paper space is not to be used. Each printed drawing sheet must be represented by a unique CAD file.

3.3 Presentation

3.3.1 #Drawing sheets

Choose an option. *See* Fig 3.2 for an example of a drawing sheet.

3.3 Presentation

3.3.1 Drawing sheets

All hard-copy output is to be printed on ISO A size standard sheets.

Either:

(a) the client will supply the contractor with a standard sheet format (including a standard north arrow, scale bar and rubric), as a CAD file, that must be used for all plotted sheets; or

(b) the contractor is to prepare a suitable sheet format for approval by the client. *See* Fig 3.2 for an example of a drawing sheet.

Where hard copy is required, each formatted CAD drawing sheet or paper space layout is to be printed.

3.3.2 #Standard views

Choose an option.

3.3.2 Standard views

Elevations and sectional elevations

Each subject to be surveyed is to be presented as an orthogonal view and, as far as is practical, all data should be presented 'square-on' to the plotted sheet.

Where an elevation or image extends over more than one sheet either:

(a) an overlap between sheets of at least 0.5m in reality is required; or

(b) the detail is to be butt jointed.

Small registration crosses are to be printed on each sheet so that adjacent sheets can be accurately aligned.

Measured building plans

Plans are to be orientated so that north is towards the top of the sheet or the principal axes of the building are parallel to the sheet edges. Where possible, the grid should be parallel to the sheet edges. If a skewed grid is unavoidable to fit the subject logically on the sheet, then text associated with the grid must be on the same alignment as the grid with all other text aligned parallel to the sheet edge.

Topographic surveys

Topographic surveys either:

(a) must be orientated so that north is at the top of the sheets; or

(b) may be orientated so that the detail fits the sheet and the grid is skewed.

For all measured building plans and topographic surveys, each drawing sheet must have a north arrow. The north arrow should not clash with any detail.

3.3.3 Layout

The following project specific information is to be included within the standard sheet layout.

- Each area surveyed is to be named correctly on the drawing sheet with reference to the actual orientation of the historic building or monument.

- Sub-titles should be placed to the bottom left of the subject where possible so that there is no risk of a title being shared by two different views. Sub-titles such as 'section at AA looking west' must refer to a clearly marked key plan or accompanying plan sheet.

- A location diagram is to be included in the title box of each drawing sheet. The diagram may be schematic if necessary and should be easily understood by a third party.

- The name of the CAD file, used to generate each sheet is to be included in the title box (*see* section 3.1.1).

- A grid for plans and topographic surveys is to be shown as intersection points or rapier marks on the sheet edge.

- Level ticks for elevations, related to the site datum, are to be placed down each side of the drawing sheet.

- A north arrow, scale bar, height datum description or reference and a key to any abbreviations used are to be included.

- Where detail extends over more than one sheet a reference diagram of the sheet layout is required (Fig 3.2).

3.3.4 Numbering of sheets

Each sheet of the survey should have a unique reference number starting at 1.

3.3.5 Data sheet for measured building and topographic survey

One sheet is to be the data sheet and must contain the following details:

- a traverse diagram;

- a sheet layout diagram;

- witness diagrams for permanently marked points;

- the listing of coordinates for all traverse stations in eastings (X), northings (Y) and height (Z) to three decimal places; and

- a full description of height data including benchmarks, where used, with the levels to three decimal places.

3.3.6 #Plotting of drawing sheets

Insert the required scale.

Choose an option. Images will appear brighter and clearer when plotted on glossy paper but may not be so suitable for use on site. Images printed on film are usually more robust and waterproof.

3.3.6 Plotting of drawing sheets

Scale

The required scale of the plotted drawing sheets is 1: ......

Accuracy

Drawing sheets must be plotted with sufficient accuracy so that any distance measured in either axis of the format is within 0.5mm of the true value at plot size.

Media and resolution

Each drawing sheet supplied as the final submission of the survey is to be of archival quality and must meet the following criteria:

- The drawing is to be plotted onto stable double-matt polyester film of at least 75 microns thickness.

- The drawing is to be plotted using an inkjet or similar plotter that has a minimum output resolution of 600dpi.

- The substrate must be compatible for use with traditional drafting inks.

- The ink is to be printed on the top surface of the film.

Rectified photography and orthophotography are to be printed either:

(a) on photographic-quality glossy paper of at least 90gsm (grams per square metre); or

(b) as described above.

The proposed output device, media and resolution are to be included in the method statement.

PDF files

Where PDF versions of drawings are required, these must meet the following criteria:

- The drawing is to be at the correct scale, as noted in the title box, when printed at 100 per cent.

- All line work is to be black except for logos etc in the title box, which should be their true colour.

- Each drawing sheet is to be represented by a separate PDF file, even when derived from multiple layout tabs in the same CAD drawing.

- Each file is to be named with the same file name as the CAD drawing file and with a suitable suffix if derived from multiple layout tabs eg DOV14T01(01) etc.

3.3.7 #Printing of rectified photography and ortho-images

Choose an option.

3.3.7 Printing of rectified photography and ortho-images

Rectified photography and ortho-imagery is required in the following format(s).

(a) image montages inserted and correctly georeferenced into a standard CAD drawing sheet and printed to the required scale; or

(b) separate images printed to the required scale, on ISO A4 or A3 photographic-quality paper. A margin of 10mm is to be left around the perimeter of each image, widened to 20mm on the left-hand edge to allow for hole punching; or

(c) other (specify).

3.4 Survey report

A survey report is to be supplied containing a brief description of the project, plus the following:

- a traverse diagram;

- witness diagrams for permanently marked points; and

- a listing of coordinates of all traverse stations and control points in eastings (X), northings (Y) and height (Z) to three decimal places,

and where applicable:

- photo location diagram;

- control prints;

- target location diagrams; and

- calibration certificates.

3.4.1 Photo location diagram

Where applicable, a sketch diagram is required to show the position from which each photograph was taken. This may be supplied as a PDF file.

3.4.2 Control prints

Where applicable, control prints are to be marked in red with the control point positions and numbers. Any detail points used are to be clearly annotated or described with a separate sketch diagram. All control prints are to be appropriately titled including the respective photo numbers. The 'prints' may be supplied in digital form only as a PDF file.

3.4.3 Imagery metadata

Camera type and other information should appear automatically in the EXIF fields. The correct date and time should be set.

Further fields may be included as necessary.

3.4.3 Imagery metadata

Each image file is to include the following minimum level of metadata in the appropriate IPTC and EXIF fields

Field	Comments
Site name	
Part of site	
List entry number	where applicable
Project reference number	or job number (*see* section 1.10.1)
Contractor's name	
Image number	each image must have a unique number
Camera type	make and model
Lens etc	
Exposure information	for original images
Date	photograph captured or image created

3.4.4 Survey material to be supplied

As well as the final CAD files, attached image files, PDFs etc, the following material is to be provided separately:

1   the original digital images in the formats as specified in section 3.1.3; and

2   the original and registered laser scan files as specified in section 7.3.1.

3.5 Provision of survey material

3.5.1 #Samples

Choose an option. It may be helpful to require samples for large or complicated projects or from untested contractors.

3.5.3 #Delivery of material

Insert delivery requirements and address etc.

3.5.4 #Transfer medium

Choose an option. It will be possible to supply some types of survey by e-mail or ftp. Where large numbers of digital images or laser scan files are involved it may be easier to use a portable hard disk.

3.5 Provision of survey material

3.5.1 Samples

Either:

(a) an initial sample of the survey in the form of a PDF file and CAD file (with attached image files as appropriate) is to be provided to the client for approval before the rest of the survey is processed; or

(b) samples are not required.

3.5.2 Preliminary plots or PDFs

Before the final plots on film or photographic quality paper are despatched, a complete set of preliminary plots on plain paper, or as PDFs, is to be supplied to the client. Plotting of the final drawing sheets should only commence after these preliminary plots or PDFs have been approved. *See* section 3.3.6 for PDF standards.

3.5.3 Delivery of material

The survey material is to be delivered to the following addresses

Name and address 1

1× set of preliminary plots on paper for approval before final delivery

1× set of final plots on film

1× set of digital data

Plus all deliverables described in section 3.4.4 as applicable.

Name and address 2

1× set of final plots on paper

1× set of final plots on film

1× set of digital data

3.5.4 Transfer medium

All of the required digital data, survey report and associated listings (eg the results of photogrammetric orientation or laser scan registration) are to be supplied on:

(a) CD-ROM; and/or

(b) DVD-R; and/or

(c) a portable hard drive or solid state/flash drive; and/or

(d) other (specify).

All disks are to be suitably labelled with the site name, date and survey reference number if provided (*see* section 1.1.1).

Appendix 3.1 #Abbreviations for survey annotation

This is not an exhaustive list.

Abbreviations must be listed on the index sheet or in the title panel. New abbreviations may be created but must be consistent within a survey or surveys. Where abbreviation leads to ambiguity, the full text is to be used.

All abbreviations used must be positioned as specified in sections 5.5.7 and 6.6.5.

Appendix 3.1 Abbreviations for survey annotation

word	abbreviation	word	abbreviation
aggregate	Agg	lath and plaster	L&P
air brick	AB	lead	Pb
aluminium	Al	manhole	MH
approximate	approx	OS bench mark	OSBM
arch height	AH	overhead	O/H
asbestos	Asb	petrol interceptor	PI
asphalt	Ap	radiator	Rad
beam height	BH	radius	Rd (state units)
bitumen	Bit	rainwater hopper	RWH
brickwork	Bk	rainwater pipe	RWP
cast iron	CI	recessed doormat	RDM
ceiling	C	reinforced concrete	RC
centre line	C	rising main	RM
cement	Cem	rain water outlet	RWO
clearing eye	CE	rain water pipe	RWP
concrete	Conc	rodding eye	RE
corrugated	Corr	round	Rd
cover level	CL	sill	S
cupboard	Cup	site bench mark	SBM
diameter	Dia (state units)	skirting board	Skrtg
downpipe	DP	soffit	Soff
drain	Dr	soil pipe	SP
drinking fountain	DF	soil and vent pipe	SVP
drive shaft	D shft	springing line	SL
earth closet	EC	stair	Str
earth rod	ER	stand pipe	St.P
electricity	Elec	stone	St
fire hydrant	FH	stop valve	SV
fireplace	FP	street gully	SG
floor	Flr	surface level	SfceL
grease trap	GT	survey station	STN or STA
ground level	GL	temporary bench mark	TBM
gully	G	void	Vd
height	Ht	volume	Vol (state units)
high level	HL	void	Vd
inspection cover	IC	wall	W
interception trap	IT	wash hand basin	WHB
internal	Int	waste pipe	WP
invert	Inv	water closet	WC
invert level	IL	width	W (state units)
lamp post	LP	window head height	WH
lightning conductor	LC	yard gully	YG

Fig 3.1 CAD coordinate systems – WCS and a UCS

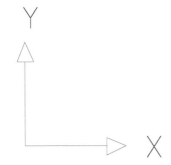

User Coordinate System (UCS) – called FRONT

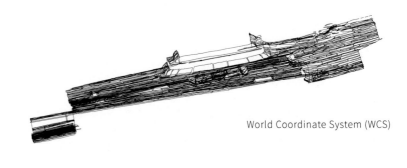

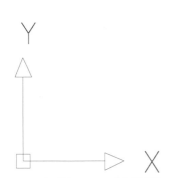

World Coordinate System (WCS)

Fig 3.2 Standard drawing sheet format (with plan aligned to border)

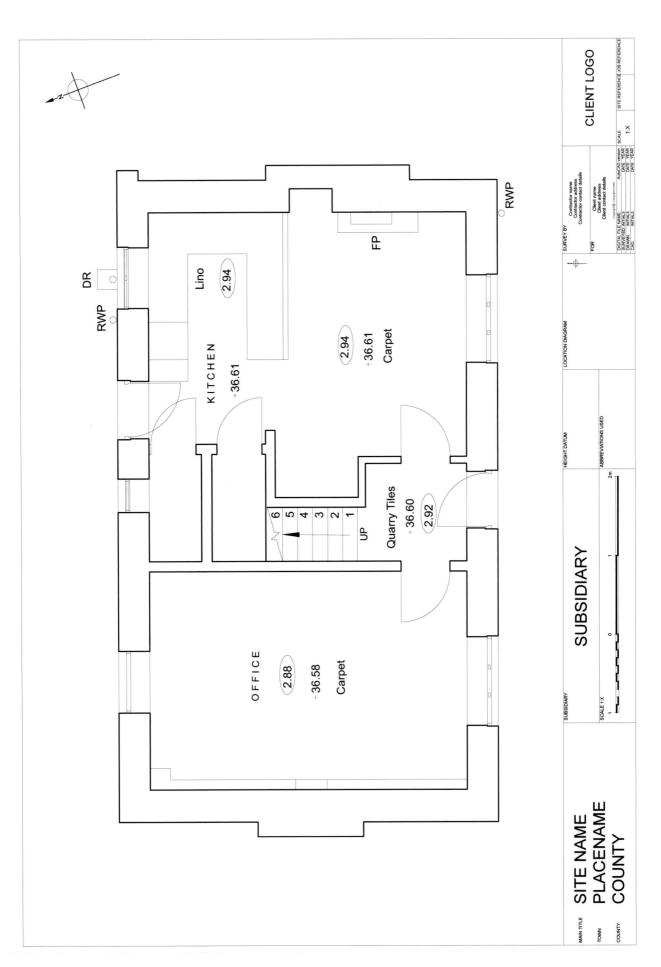

Standard specification for image-based survey

4.1 Image-based survey techniques

4.1.1 Photogrammetric survey

Photogrammetry is 'the science, and art, of determining the size and shape of objects as a consequence of analysing images recorded on film or electronic media'.

KB Atkinson, 1996 *Close Range Photogrammetry and Machine Vision*

This definition could be regarded as encompassing all image-based measurement techniques, including structure from motion. Although scaled outline drawings can be generated using digitising techniques to record common points from overlapping images viewed monoscopically, this technique is not covered in this document. If a contractor wishes to use this technique to provide drawings he/she must be able to demonstrate that the performance criteria for 3-D data, set out in section 4.5, have been fully satisfied and that the specified standards for drawing detail and presentation are met.

Structure from motion and multi-view stereo techniques

The combination of structure from motion (SfM) and multi-view stereo is a photogrammetric technique derived in part from computer vision, wherein a series of overlapping but uncalibrated input images permit the interior and exterior orientations of the camera(s) used to be estimated and a model derived. Accuracy can be greatly improved by the use of calibrated camera/lens set-ups and the introduction of rigorous control networks, although the requirements are lower than those typically used for more traditional photogrammetric processing.

4.1.2 Orthophotographic survey

An orthophotograph is defined as a digital image that has been corrected for scale errors due to tilt and depth displacement. Each pixel will have been individually scaled and shifted in order to produce an orthographic projection, as opposed to the perspective projection of ordinary photographs (Figs 4.3 and 4.4). This is normally achieved using software that requires a digital surface model (DSM) to calculate the necessary scaling and translations. The DSM may be created photogrammetrically, usually in the same software, or could be imported from another source such as laser scanning. A number of images may be corrected and montaged together to provide the required coverage.

4.1.3 Rectified photographic survey

Rectified photographic surveys are defined as those where single photographs are taken with the image plane of the camera approximately parallel to the principal plane of the object and then further digitally corrected so that scale errors due to camera tilt, but not those due to relief, have been removed by means of a projective transformation. The usual product of the technique is a scaled photographic image or a montage of a number of images (Figs 4.5, 4.6 and 4.7).

4.1 Image-based survey techniques

Image-based surveys are defined as surveys where photographic images, together with an element of scale, have been used to generate the required detail, presented in either a line drawing or scaled image format.

4.1.1 Photogrammetric survey

For the purposes of this document photogrammetric surveys are defined as surveys where overlapping image sets are used together with control to produce a three-dimensional representation of the subject from which the required detail is generated. The traditional product is a scaled outline drawing (Figs 4.1 and 4.2). It is also possible to present detail in a scaled image format, such as an orthophotograph (Figs 4.3 and 4.4).

For the sake of brevity the combined use of structure from motion and multi-view stereo techniques will be referred to as SfM from this point on.

4.1.2 Orthophotographic survey

For the purposes of this specification an orthophotograph is defined as a digital image that has been corrected for scale errors due to both camera tilt and depth displacement.

4.1.3 Rectified photographic survey

Rectified photographic surveys are defined as those where single photographs are taken with the image plane of the camera approximately parallel to the principal plane of the object and then further digitally corrected to remove scale errors due to camera tilt.

4.2 Control for image-based surveys

4.2.1 #Accuracy

Choose an option. Option (a) ±3mm is sufficient for the standard architectural scales.

4.2.2 Control of subject

Where SfM is used the term 'model' is taken to mean the result of processing any number of images together. Whilst the specified four control points will be sufficient to scale and translate the model into the required coordinate system, many more may be needed to achieve the internal accuracy required to satisfy the clauses of section 4.5.

4.2.3 #Use of detail points

Choose an option. Access problems may mean it is necessary to use solely detail points on upper levels. Some historic fabric may be too fragile for the application of targets (*see* section 2.2.9).

4.2 Control for image-based surveys

4.2.1 Accuracy

Image control points are to be provided to a 3-D accuracy of either:

(a) ±3mm; or

(b) other (specify).

A listing of the 3-D coordinates is to be included in the survey report.

4.2.2 Control of subject

For photogrammetric and orthophotographic surveys a minimum of four coordinated control points, directly observed in the field, are to be provided for each model. Where practicable, targets are to be placed on the fabric (*see* sections 1.6.2 and 2.2.9) and must:

- be no larger than 60mm × 40mm;

- no thicker than 0.5mm; and

- have a matt, non-reflective surface finish.

4.2.3 Use of detail points

Where targets cannot be placed on the fabric it is acceptable to use unambiguous points of detail. A sketch diagram or annotated image showing the location of each point is to be included in the survey report. Detail points must be easily identifiable and must not be taken from the extreme edges of the subject. It will:

(a) not be acceptable; or

(b) only be acceptable where absolutely essential; or

(c) be necessary

solely to use detail points in an image or model.

4.3 Photogrammetric and orthophotographic surveys

4.3.1 #Digital cameras

While the ground sample distance (gsd) specified in section 4.4.2 could be achieved using a digital camera with an array of less than 13 million pixels, this would, in most cases, result in uneconomic coverage as the photographs would have to be taken closer to the subject. A camera with a pixel size of less than 6 microns could also achieve the required gsd, but this would be at the expense of increased image noise – that is, unwanted variations in brightness and colour information not present in the subject.

Standard digital cameras can be calibrated for photogrammetry as long as some way is found to fix the focus of the lens so that a precise focal length or principal distance can be measured. SfM systems will benefit from the use of a calibrated camera, but it is not absolutely essential as self-calibration can be employed.

For production of colour orthophotographs good colour balance is essential. However, photographs for close-range applications will, unlike aerial photography, usually have been taken under varying exposure conditions. This means that white balance will need to adjusted using a suitable colour chart or reference card.

Pragmatism means that the requirement for pixel size must be relaxed for fixed-wing, small unmanned aircraft (SUA) as there are few currently available that have sufficient payload to carry the type of cameras that meet the specification. The convenience of acquiring aerial photography in this way must be weighed against possible lower quality due to increased image noise.

4.3.2 Digital image criteria

Monochrome images will have a smaller file size and by definition avoid colour balance problems. Colour images are ubiquitous, may assist in the interpretation of detail and will be essential for recording wall paintings, mosaics, tiles etc.

4.3 Photogrammetric and orthophotographic surveys

4.3.1 Digital cameras

The following criteria must be fulfilled:

- Digital cameras must have a sensor array with at least 13 million pixels and each pixel must a have a minimum size of 6 microns.

- Digital cameras must have a fixed-focus lens with minimal distortion. This is to be calibrated to provide a precise focal length measured to within 0.01mm, and the precise distortion characteristics measured to enable compensation to occur during processing. A copy of the calibration certificate is to be supplied with the final survey materials. Details of proposed cameras and lenses are to be included in the method statement for each survey.

- For SfM techniques use of a fixed focus calibrated lens is either:

    (a) essential; or

    (b) not essential.

Aerial photography using fixed-wing, small unmanned aircraft (SUA) may be acquired using compact cameras as long as the requirements for ground sample distance (gsd), as set out in section 4.4.2, are met.

4.3.2 Digital image criteria

Digital imagery must be captured as 16-bit but is to be reduced to 8-bit for processing. Images are to be captured in RAW format and these files must be supplied as well as TIFF versions. Monochrome imagery may be provided by reduction to greyscale in a standard image processing package. Colour imagery, as well as meeting the above criteria, is to be balanced for either daylight or artificial illumination as appropriate. The required colour space is the Adobe RGB (1998) ICC colour profile. A standard colour chart and/or greyscale is to appear in at least one of the images per subject area to provide guidance on colour balancing prior to output.

4.4 Image acquisition for photogrammetry and orthophotography

4.4.1 Imagery arrangement

The base to subject distance ratio (known as base to height ratio for aerial photography) is the ratio of the distance between two camera positions and their distance from the subject. If the ratio is too low or too high it will not be possible to view the photographs stereoscopically. An overlap of 60 per cent between photographs ensures complete but economic coverage from a strip of photographs. Excessive variations in scale between stereo images will make them difficult to view stereoscopically.

4.4.2 #Ground sample distance

State the required ground sample distance (gsd). Gsd is the size in the real world of that part of the subject represented by one pixel of a digital image. It is a function of focal length, camera to subject distance (or flying height) and pixel size.

$$gsd = (H/f) \times p$$

where

H = camera to subject distance or flying height

f = focal length

p = pixel size (sensor size in one axis divided by pixel count in same axis)

For photogrammetry at the typical architectural scales the following values are recommended:

for 1:50 output scale, 3mm maximum gsd

for 1:20 output scale, 2mm maximum gsd

for 1:10 output scale, 1mm maximum gsd

For topographic survey or orthophotographs from SUA aerial photography the following values are recommended

for 1:500 output scale, 4cm maximum gsd

for 1:200 output scale, 2cm maximum gsd

for 1:100 output scale, 1cm maximum gsd

Values for different output scales may be extrapolated.

4.4 Image acquisition for photogrammetry and orthophotography

The whole of the subject area must be covered by overlapping imagery.

4.4.1 Imagery arrangement

Images must be arranged to provide the following basic geometry and camera alignments:

- camera base to subject distance ratio of no more than 1:4;

- overlap between adjacent stereo images of at least 60 per cent (80 per cent for SfM);

- overlap between adjacent strips of stereo images of at least 10 per cent (40 per cent for SfM).

The following must also be observed, except where SfM techniques will be employed:

- alignment of each image plane, with the principal plane of the subject, to be within ±3° of parallelism;

- minimised vertical tilt of the camera, either upwards or downwards, to a maximum of 15°;

- a variation in the scale between adjacent stereo-images of no more than 5 per cent.

4.4.2 Ground sample distance

The ground sample distance (gsd) for each image is to be a maximum of ......mm.

4.4.3 Coverage of prominent architectural features

Where prominent architectural features are present, such as a large window or arched doorway, imagery must be taken that provides an orthogonal and not an oblique or tilted view of the feature. This is particularly important when the imagery is to be used to form a scaled image, such as an orthophotograph. Where SfM techniques are to be employed additional oblique imagery may be desirable to increase achievable coverage.

4.4.4 Use of oblique imagery

In general photography is to be taken as square-on to the subject as practicable. Oblique imagery may be needed to infill areas potentially obscured on the standard orthogonal imagery or to provide suitable low-level aerial coverage from an SUA platform. Unless SfM techniques are to be employed the imagery should not be convergent and the camera axes must not cross.

4.4.5 #High-level coverage

Choose an option. Photography of high elevations or even lower elevations with reduced stand-off distance can suffer from occlusions caused by the relief of the detail. This can lead to gaps in the plotted detail or orthophotograph. Extreme tilts of the camera can result in stereo-models that will not set up or are difficult to plot from. To avoid these problems the camera can be raised up using access equipment such as a scaffold tower, mast, hydraulic lift or SUA.

4.4.6 Use of small unmanned aircraft (SUA)

Choose the required option(s). Many different names and acronyms are used to describe the low-level operation of unmanned, remotely-controlled aerial platforms that can be used to acquire imagery suitable for survey applications. These include:

- drone (usually in a military context);

- remotely piloted vehicle (RPV);

- remotely piloted aircraft system (RPAS);

- unmanned aircraft system (UAS);

- unmanned aerial vehicle (UAV); and

- small unmanned aircraft (SUA) – the term used by the UK Civil Aviation Authority (CAA) that is referred to within their latest regulatory information (**www.caa.co.uk/ in2014081**).

Balloon- and kite-based aerial platforms do not require CAA permission to fly below 60m. For the purpose of this document, however, all reference to SUA should be taken to apply to all alternative names and acronyms for remotely-controlled, low-level aerial platforms.

SUAs can be of either fixed-wing or rotary-wing design. Both are able to carry, and enable the remote operation of, a digital sensor for the capture of low-level digital imagery. Typical sensors include still and video cameras, both compact, DSLR and purpose-designed units. Once captured the imagery has a number of heritage-related applications. These include condition monitoring, asset inventory, infrastructure modelling, volumetric analysis, presentation, multimedia products, filming and journalism. If captured in an overlapping form, whether stereo or multi-image, the imagery can also be used to photogrammetrically derive three-dimensional survey data of buildings and landscapes.

Current CAA regulations allow a SUA (including payload) of up to 20kg weight to be flown within UK airspace. However, they are restricted to a flying height not exceeding 400ft/122m above ground level and a distance not beyond the visual range of the operator up to a maximum distance of 500m (IN-2014/081, CAA, 2014). Although typically less than 2m in size, some SUAs can achieve speeds of up to 50mph, meaning that when in flight they possess significant momentum and potential to cause damage if not used in a safe and controlled manner. Further guidance on these requirements can be found in the CAA document CAP 722 Unmanned Aircraft System Operations in UK Airspace – Guidance within the sections that relate to commercial flights using small unmanned aircraft, and also at the CAA website **www.caa.co.uk/uas**.

4.4.5 High-level coverage

Where the subject to be surveyed is of a significant height, imagery must still be taken within the stated tolerances for camera tilt, image scale variation and gsd as outlined in sections 4.4.1 and 4.4.2. The use of access equipment is:

(a) not essential; or

(b) at the contractor's discretion; or

(c) essential.

4.4.6 Use of small unmanned aircraft (SUA)

All image acquisition undertaken with a SUA platform must conform to current UK Aviation Law, as detailed in the Air Navigation Order (specifically articles 138, 166 and 167). A current and valid CAA 'Permission for Aerial Work' and evidence of public liability insurance cover must be supplied prior to survey commencing. The conditions listed within each individual permission must be specifically checked to confirm they are appropriate for the specified survey task.

Full details of any proposed use of a SUA platform must be included in the method statement and include both a flight plan, showing the proposed area and flying route, and a risk assessment that considers all associated risks and how these will mitigated.

Where overlapping SUA imagery is required for survey applications the ground sample distance (gsd) is to be either:

(a) as specified in 4.4.2; or

(b) a maximum of ......mm.

Where vertical aerial imagery is required for SfM or other photogrammetric applications, the overlap between adjacent:

• stereo images is to be at least 80 per cent; and

• strips of images is to be at least 40 per cent.

Oblique imagery at approximately 15° off-nadir to brace the SfM model will either:

(a) be required; or

(b) not be required.

Where non-overlapping SUA imagery is also required for pictorial applications, it is to be captured either:

(a) at a flying height of ......m and from the following orientations (specify); or

(b) as described in the attached photography brief.

4.4.7 #Completeness of survey

Choose an option. Field completion will add to initial costs but may reduce those of subsequent work, such as archaeological analysis.

4.4.7 Completeness of survey

Survey data obtained using image-based techniques is to be as complete as possible. The client will endeavour to provide a clear and unobstructed view for photography prior to survey commencing, but where obstructions prevent the use of images it will be acceptable to omit detail that cannot be clearly seen (*see* section 4.6.2). Field completion by another method will either:

(a) not be required; or

(b) be required. Describe the proposed technique in the method statement.

4.5 Photogrammetric processing

4.5.1 Accuracy of orientation

Photogrammetric orientation of overlapping photographs is required before they can be used for accurate survey and consists of three processes. These are the interior, relative and absolute orientation (the latter two are often performed together and described as the exterior orientation). The interior orientation accounts for the geometry of the camera. The relative orientation recreates the positions and tilts of the camera relative to each other when the photographs were taken. The absolute orientation uses control points to position the stereo view in 3-D space so that scaled detail in the correct location can be recorded. Any discrepancies between the coordinates of the control points in the model and their true coordinates as measured in the field are displayed as residuals. In order to achieve the required accuracy of processing, the residuals will have to be equal to or less than the figures stated in section 4.5.2.

4.5 Photogrammetric processing

All photogrammetric processing work is to be carried out using dedicated photogrammetric software utilising overlapping imagery. The choice of equipment and methodology is discretionary, but must be outlined in the method statement. Material generated must be within the stated tolerances and meet the specified standard for drawing detail and presentation. The data may be required in both digital form and as a hard copy (*see* section 3.5 for details).

4.5.1 Accuracy of orientation

All overlapping images are to be processed, so that the residuals obtained during the orientation procedure enable the generation of survey data that is commensurate with the line width accuracy at final plot scale. The orientation results for all processed models are to be recorded and provided, as a digital listing, with the final survey materials.

4.5.2 Accuracy of processing

For the production of line drawings, recorded points must be within the accuracy figures noted below, the standard for photogrammetric processing relates to the accuracy of final line width of the vector data generated. For output at standard architectural scales, using a 0.18mm line width these are:

> for 1:50 output scale, 9mm in reality
>
> for 1:20 output scale, 4mm in reality
>
> for 1:10 output scale, 2mm in reality

and for standard topographic scales, using a 0.18mm line width:

> for 1:500 output scale, 90mm in reality
>
> for 1:200 output scale, 40mm in reality
>
> for 1:100 output scale, 20mm in reality

4.5.3 #Digitising

Choose the option required and insert suitable values if necessary. The default values will be suitable in most cases (*see* section 2.1.2).

4.5.5 #Output scale

Insert the required output scale.

4.5.3 Digitising

Where regular un-eroded features are apparent, both stream and point-by-point digitising methods may be used to accurately transcribe the shape of the feature being surveyed. For any irregular, eroded features the following criteria must be fulfilled:

- Stream digitising methods must be used with a maximum distance between points of either:

  (a) 30mm; or

  (b) …mm in reality.

- Points must be recorded, whether manually or automatically, at distinct corners and changes in direction of greater than either:

  (a) 10°; or

  (b) ……

See section 2.1.2 for further details.

4.5.4 3-D data

All photogrammetric line work is to be recorded as 3-D CAD data. Features such as the splayed reveals of window openings, the curved elements of moulding profiles and returns to door openings must be correctly recorded in 3-D so as to allow oblique viewing of the final dataset. Care is to be taken to ensure that no unnecessary overlap of lines in 3-D space occurs.

4.5.5 Output scale

The final output scale is to be 1: …

4.6 Drawing content

4.6.1 #Level of detail

Choose the required level of detail.

Delete any items not required. The use of CAD repeats is unlikely to be appropriate for historic buildings as even detail that appears to be identical can have subtle differences.

4.6 Drawing content

4.6.1 Level of detail

The areas identified for survey will require photogrammetric processing to either:

(a) the full level of detail; or

(b) the outline level of detail, as described below.

Full detail

All architectural detail is to be recorded, including:

- windows;

- doors;

- fireplaces;

- jambs, sills, string courses, lintels;

- window tracery and ferramenta (ironwork);

- architectural fragments including corbels, architraves and mouldings;

- roof and chimney outline;

- any visible cracks in fabric;

- quoins and individual voussoirs above window openings;

- services and rainwater goods;

- changes of surface treatment and images upon fabric, eg wall painting;

- all visible ashlar, cut or dressed stone and coursed rubble; and

- revealed core-work (outline of medium- to large-sized stones)

(*see* Fig 4.1)

The outline of individual bricks and designs in stained glass windows are not normally required. Where the jointing between stonework is smaller than the required tolerance of the survey, and cannot be recorded by drawing the outline of each block, a single line is to be placed along the centre line of the joint.

Outline detail

Where an outline survey is specified, only the principal architectural detail is to be recorded. Unless specified, each individual stone or brick is not to be recorded.

The maximum output scale of survey dictates the required accuracy of photogrammetric processing. Therefore any feature(s) that measure greater or equal to the figures noted in section 4.5.2 must be recorded by a solid line, if visible in the model. Where a feature is not visible in the model, it must not be recorded.

The use of CAD repeats or cloning of features is either:

(a) not permitted; or

(b) permitted.

(*see* Fig 4.2).

4.6.2 #Specific details to be noted

State the required level of recording for sculptural detail.

Choose whether individual roof tiles are required.

State the required level of recording for any vaulting details, timber panelling or metal work.

Insert any other required details or delete those not required.

4.6.2 Specific details to be noted

The outline of any stone block shall be taken as the arris where this is visible. Otherwise the junction between stone and mortar is to be recorded.

- Features in stonework such as putlog holes, structural cracks and rainwater services must be recorded in full.

- Window reveals must be recorded in full (Fig 4.10; *see* section 4.6.6 for further details on presentation).

- The outline of any areas that cannot be surveyed using photographic-based survey must be plotted as a dashed line encompassing the text 'obscured by ......'. For example, when detail is hidden by vegetation, the text 'obscured by vegetation' is to be inserted.

- The side faces of any buttresses that are more than one stone in depth are to be surveyed and presented as separate orthogonal drawings.

- The required level of recording of sculptural detail is ......

- Individual roof tiles will either:

  (a) not be required; or

  (b) be required.

- The required level of any vaulting details, timber panelling and metalwork is ......

- Where the edge of an individual feature, such as a stone or brick, forms part of the outer edge of the elevation being surveyed, this must not be recorded separately as part of a major outline. Each object must be recorded as a closed feature with a separate line in between to represent any mortar infill (*see* Fig 4.8).

- Other details ......

4.6.3 Line styles

The standard line type for all processed architectural detail will be a continuous solid black line of 0.18mm in width. A dashed line is to be used to accurately define the shape where:

- the edge of a feature is eroded/weathered beyond a range of 20mm;

- the edge of a feature is indistinct and the operator cannot guarantee the accuracy of the line work;

- the limits of an area of obscured detail (eg by vegetation are required)

(*see* Fig 4.9).

It is acceptable to use arcs, so long as they define accurately the shape of the feature to be surveyed.

4.6.4 #Curved features

Choose the required option. Unwrapping will allow the accurate scaling of dimensions and areas from a drawing, but the detail will be displaced from its true 3-D position. An orthogonal view of a curved feature will suffer from foreshortening, but the detail will be in the correct position.

4.6.5 Closed features

Plotting stones etc as closed polylines means it is possible to apply hatching etc in the CAD drawing.

4.6.7 #Provision of sectional information

The photogrammetric process allows the production of horizontal (profile) and vertical (section) cut lines through the subject being surveyed. The standard level for horizontal profiles to be taken is 0.1m above window sill level, although the precise location and purpose should be noted on an attachment. Unless they are to form part of an architectural section, these cut lines should be presented as a single, continuous solid line.

Indicate whether any sectional information is required and if so provide a diagram or description to indicate the location of cut lines.

4.6.4 Curved features

Curved features are to be recorded in true 3-D, presented either:

  (a) unwrapped so as to provide a true-to-scale representation; or

  (b) as an orthogonal view.

The method proposed for any required unwrapping of data is to be outlined in the method statement.

4.6.5 Closed features

Detail that is a closed feature, such as a complete stone, is to be recorded with a closed 3-D polyline. Where a feature does not appear closed, such as part of an obscured stone, it is to be drawn as an unclosed 3-D polyline.

4.6.6 Recording of reveals

Where detail of reveals is recorded, each of the faces forming the arris is to be plotted. Problems of overlapping detail can occur when viewed in 2-D. This is particularly apparent when the reveal is at right-angles to the main face. To avoid this, detail that is obscured by the main elevation is to be placed in a separate CAD layer, eg 0P-opening_A (*see* Fig 4.10).

4.6.7 Provision of sectional information

Sectional information is either:

  (a) not required; or

  (b) required for the cut lines shown on the attached diagram.

4.6.8 #CAD layer names

The Historic England convention may be used as a default or any other convention may be substituted. Please note that the convention separates detail by function rather than form.

4.6.9 #CAD layering – general notes

If a convention other than the Historic England standard is used, this section will need to be edited.

4.6.8 CAD layer names

The following is the Historic England convention for the layering of architectural photogrammetric survey in CAD (Fig 4.9). Please note that the prefix is 0(zero)P not OP.

layer	colour	description
0P-major	white	major – this is to include all structural elements facing stone, ashlar etc, except for those specified below
0P-core	red	core-work exposed by the removal of facing stone
0P-openings	blue	windows/doors/fireplaces – this is to include all jambs, sills, voussoirs, lintels and surrounding stonework
0P-architectural	green	architectural fragments: corbels, architrave, mouldings, etc
0P-sculptural	cyan	sculptural detail: figures and carved detail
0P-services	magenta	modern service: drainpipes, lightning conductors, ducting, etc
0P-text	white	text/notes: for areas obscured, relating to architectural data, not border information
0P-control	white	control points: depicted as a cross with point number; layer to be frozen during hard copy output
0P-hidden	grey	hidden detail
0P-etc		these may be used where a particular element does not fit into the previous layers; the layer name is to be prefixed 0P-

4.6.9 CAD layering – general notes

Where architectural fragments or sculptural features form part of a window or door, etc, they are to appear within the layer for windows/doors (eg 0P-openings). The above layering convention is also to be applied when section or profile information is specified, depending upon the type of detail that the cut line actually passes through. Any areas of erosion or damage that are recorded should be placed within the same layer as the feature that they concern. If there is any doubt into which layer a feature should be placed, it should be put into 0P-major (see Fig 4.9).

4.7 Orthophotographic processing

Please note this section specifies close range orthophotographic processing. For a conventional aerial orthophotograph specification, refer to RICS, 2010 *Vertical Aerial Photography and Derived Digital Imagery*, 5 edn (RICS Guidance Note) (GN 61/2010).

Orthophotographic processing work may be carried out using a standard photogrammetric workstation utilising stereo-imagery or SfM software.

4.7.1 Accuracy of orientation

See notes to section 4.5.1.

4.7.2 #Digital elevation model

The digital elevation model (DEM) required for the production of an orthophotograph may be generated photogrammetrically or from a laser scan point cloud. Choose the option required and insert a suitable value if necessary. The values shown in option (a) will be suitable for elevations with average relief, which are to be presented at 1:10, 1:20 or 1:50 scale. Larger scales and more complicated relief will require smaller values for point spacing.

Where photogrammetric techniques are used to produce the DEM it will be the equivalent of a digital surface model (DSM).

4.7 Orthophotographic processing

The choice of equipment, software and method for providing the required survey are discretionary, but they must be outlined in the project method statement.

4.7.1 Accuracy of orientation

All overlapping images are to be processed, so that the residuals obtained during the orientation procedure enable the generation of survey data that is commensurate with the line width accuracy at final plot scale (*see* section 4.5.2). The orientation results for all processed models are to be recorded and provided, as a digital listing, with the final survey materials.

4.7.2 Digital elevation model

For stereo photogrammetric workstations the spacing of points in the digital elevation model (DEM), used for the generation of the orthophotograph, is to be either:

(a)

for 1:50 output scale, 50mm maximum

for 1:20 output scale, 20mm maximum

for 1:10 output scale, 10mm maximum

for 1:500 output scale, 500mm maximum

for 1:200 output scale, 200mm maximum

for 1:100 output scale, 100mm maximum

Break lines should be included where possible to assist in the orthophotograph generation. The processed DEM must accurately depict the 3-D surface of the subject and therefore automatically generated DEMs may require manual editing.

or

(b) where SfM methods are employed, the DEM point spacing is to be:

for 1:50 output scale, 10mm maximum

for 1:20 output scale, 5mm maximum

for 1:10 output scale, 1mm maximum

for 1:500 output scale, 40mm maximum

for 1:200 output scale, 20mm maximum

for 1:100 output scale, 10mm maximum

and break lines will not be required.

4.7.4 #Output resolution

Indicate the required scale.

The output pixel size should be as follows for the typical architectural scales:

    for 1:50 output scale, a maximum pixel size of 3mm in reality

    for 1:20 output scale, a maximum pixel size of 2mm in reality

    for 1:10 output scale, a maximum pixel size of 1mm in reality

Values for different scales may be extrapolated.

4.7.5 #Presentation of orthophotographs

Choose an option for image file format. The geoTIFF option will facilitate the importation of images into other software such as geographical information systems (GIS). The separate world file will mean that a geoTIFF file can still be geo-referenced even if the location information is deleted from the file header.

4.7.3 Mosaic generation

The orthophotograph mosaic is to be generated so that the joins between images are not visible in the final output. Seam lines should follow linear detail such as mortar joints where possible. Colour balance must be consistent and any distinct shadows in recessed areas are to be digitally removed.

4.7.4 Output resolution

The final output scale is to be 1: ......

with a maximum pixel size of ......mm in reality.

4.7.5 Presentation of orthophotographs

All orthophotographs are to be attached to an AutoCAD .DWG file and correctly geo-referenced to the control coordinate system.

The images are to be supplied as either:

    (a) TIFF files; or

    (b) geoTIFF files.

Where geoTIFF is specified, a separate world file is also required.

Wherever practicable, the joins between adjacent images of a montage should not be visible.

4.8 Rectified photography

4.8.1 Digital cameras

Most compact digital cameras now have arrays of 10 million pixels or greater; however, their sensors may be quite small and therefore subject to excessive noise – that is, unwanted variations in brightness and colour information not present in the subject. The lenses of compact cameras may well exhibit high distortion and therefore not be suitable for rectified photography. Higher-grade digital cameras are able to use a wide range of high-quality lenses that will be able to satisfy the conditions of this section.

4.8.2 Digital image criteria

Monochrome images will have a smaller file size and by definition avoid colour balance problems. Colour images are ubiquitous, may assist in the interpretation of detail and will be essential for recording wall paintings, mosaics, tiles etc.

4.8 Rectified photography

4.8.1 Digital cameras

The following criteria must be fulfilled:

- Digital cameras must have a sensor array with at least 10 million pixels and each pixel must a have a minimum size of 4 microns.

- Digital cameras must have a lens, whether calibrated or not, that exhibits minimal distortion. That is the lens must not introduce any discernible distortion of horizontal and vertical lines in the subject to be surveyed and the resultant digital images must be capable of being processed to the tolerances stated in sections 4.9.1 and 4.10.1. Alternatively lens distortion may be corrected for through post processing.

4.8.2 Digital image criteria

Digital imagery must be captured as 16-bit but is to be reduced to 8-bit for processing. Images are to be captured in RAW format and these files must be supplied as well as TIFF versions. Monochrome imagery may be provided by reduction to greyscale in a standard image processing package. Colour imagery, as well as meeting the above criteria, is to be balanced for either daylight or artificial illumination as appropriate. The required colour space is the Adobe RGB (1998) ICC colour profile. A standard colour chart and/or greyscale is to appear in at least one of the images per subject area to provide guidance on colour balancing prior to output.

4.9 Image acquisition for rectified photography

Although digital rectification packages can deal with fairly large tilts, it is always advisable to take the photography as square-on as possible in the field. This is particularly important for features such as doors and windows, as any minor relief, such as mouldings or reveals, will not be symmetrically recorded by a tilted photograph and as a result the final rectified photograph will look unnatural. Tilted images will also exhibit variation in ground sample distance (gsd). Even lighting is particularly important when a number of photographs are to be montaged together. The join between two photographs will be difficult to hide if the exposure is uneven.

4.9.1 #Ground sample distance

For digital imagery, state the required gsd. Gsd is the size in the real world of that part of the subject represented by one pixel of a digital image (see also section 4.4.2). For rectified photography at the typical architectural scales the following values are recommended:

> for 1:50 output scale, 5mm maximum gsd
>
> for 1:20 output scale, 2.5mm maximum gsd
>
> for 1:10 output scale, 1mm maximum gsd

Values for different output scales may be extrapolated.

4.9.2 #High-level coverage

Delete the options not required. Photography of high elevations or even lower elevations with reduced stand-off distance can suffer from occlusions caused by the relief of the detail. The gsd may vary significantly across the image and therefore not satisfy section 4.9.1. Extreme tilts of the camera can result in photographs that fail to rectify in standard packages. To avoid these problems the camera can be raised up using access equipment such as a scaffold tower, mast, hydraulic lift or SUA.

4.9.3 #Use of oblique imagery

Insert a figure of 45° or less. The use of oblique imagery should, however, be avoided wherever possible for the reasons stated above.

4.9 Image acquisition for rectified photography

All areas outlined for survey in the project brief are to be covered by suitable imagery. These images must be arranged to provide the following alignments and image quality:

- The alignment of each image plane, with the principal plane of the object, to within ±3° of parallelism. The alignment method is discretionary, but must be noted in the method statement provided for each survey.

- Where prominent architectural features are present, such as a large window or arched doorway, imagery must be taken that provides an orthogonal and not an oblique or tilted view of the feature.

- Imagery must be evenly lit with no strong shadows visible across the area covered.

4.9.1 Ground sample distance

The ground sample distance (gsd) for each image is to be a maximum of ......mm.

4.9.2 High-level coverage

Where the subject to be surveyed is of a significant height, imagery must still be taken within the stated tolerances for camera tilt and gsd. The use of access equipment is:

    (a) not essential; or

    (b) at the contractor's discretion; or

    (c) essential.

Should the use of a small unmanned aircraft (SUA) be proposed, refer to section 4.4.6.

4.9.3 Use of oblique imagery

This should only be used to infill areas potentially obscured on the standard orthogonal imagery or where economic coverage cannot otherwise be obtained. If used, the alignment of the image plane must be within 45° of parallelism with the principal plane of the subject. Details of where this technique is proposed are to be included in the method statement for each survey.

The maximum tilt allowed will be ......

4.9.4 #Definition of principal plane

In almost all cases, choose option (a). Elevations with multiple but well defined planes, such as a buttressed wall, can often be successfully recorded using rectified photography. In these cases careful definition of which planes are to be treated separately will be required.

4.9.5 #Control of rectified photography

Choose the required option. Option (a) coordinate-controlled rectified imagery will be the most accurate and will be almost essential if a montage of a number of different images is required, as there will need to be control points common to adjacent photographs. The control points can be acquired through the use of a total station theodolite or could be extracted from a laser scan point cloud. Option (b) scaled rectified imagery will be quicker, requires less equipment and therefore should be cheaper, but obviously will not be related to any coordinate system.

4.9.4 Definition of principal plane

The principal plane is to be either:

    (a) the largest mono-planar surface of the area to be surveyed; or

    (b) other (specify).

4.9.5 Control of rectified photography

Rectified photography is to be:

    (a) coordinate-controlled rectified imagery; and/or

    (b) scaled rectified imagery; or

    (c) controlled by a method chosen at the contractor's discretion

(*see* Figs 4.5 and 4.6).

Coordinate-controlled rectified imagery requires a reliable and repeatable coordinate system to be produced covering each subject area. Coordinates will be required for at least four survey targets or detail points if necessary, per image, to be used as 3-D control points. The method for generating the coordinate control is discretionary, although this must achieve the survey accuracies specified in section 2.1.2.

Scaled rectified imagery requires the introduction of a scale measurement in the principal plane of the subject. The following methods for providing the scale are acceptable:

- one horizontal and one vertical measured distance between targets, placed in the principal plane in each image; and

- a scale bar, typically divided into 100mm sections, placed in the principal plane and extending over at least half of the image area.

4.10 Processing rectified photography

4.10.2 #Output resolution

Indicate the required scale.

The output pixel size should be as follows for the typical architectural scales:

for 1:50 output scale, a maximum pixel size of 5mm in reality

for 1:20 output scale, a maximum pixel size of 2.5mm in reality

for 1:10 output scale, a maximum pixel size of 1mm in reality

Values for different scales may be extrapolated.

4.10.3 #Presentation of rectified photography

Choose an option. Edit option (a) if a different CAD package is required.

If it is anticipated that dimensions or detail are to be digitised from the rectified photographs then (a) will be the best option.

4.10 Processing rectified photography

The final product is to be a digital image and it is anticipated that this will be achieved using one of a number of dedicated digital rectification packages. Other methods are acceptable if it can be shown, in the method statement, that the specified tolerances can be achieved.

4.10.1 Accuracy of processing

For the typical architectural scales the required accuracies for rectified processing are:

for 1:50 output scale, 25mm in reality

for 1:20 output scale, 10mm in reality

for 1:10 output scale, 5mm in reality

4.10.2 Output resolution

The final output scale is to be 1: ......

with a maximum pixel size of ......mm in reality.

4.10.3 Presentation of rectified photography

All rectified photographs or montages are to be supplied as TIFF files. They are either:

(a) to be attached to an AutoCAD .DWG file and correctly referenced to the control coordinate system; or

(b) if single rectified images, they may be supplied as individual TIFF files, but these must be to scale when printed at 100 per cent image size.

Wherever practicable, the joins between adjacent images of a montage should not be visible. Any non-controlled planes, or surrounding detail off the principal planes of the historic building or monument, are to be cropped prior to final output (*see* Fig 4.7).

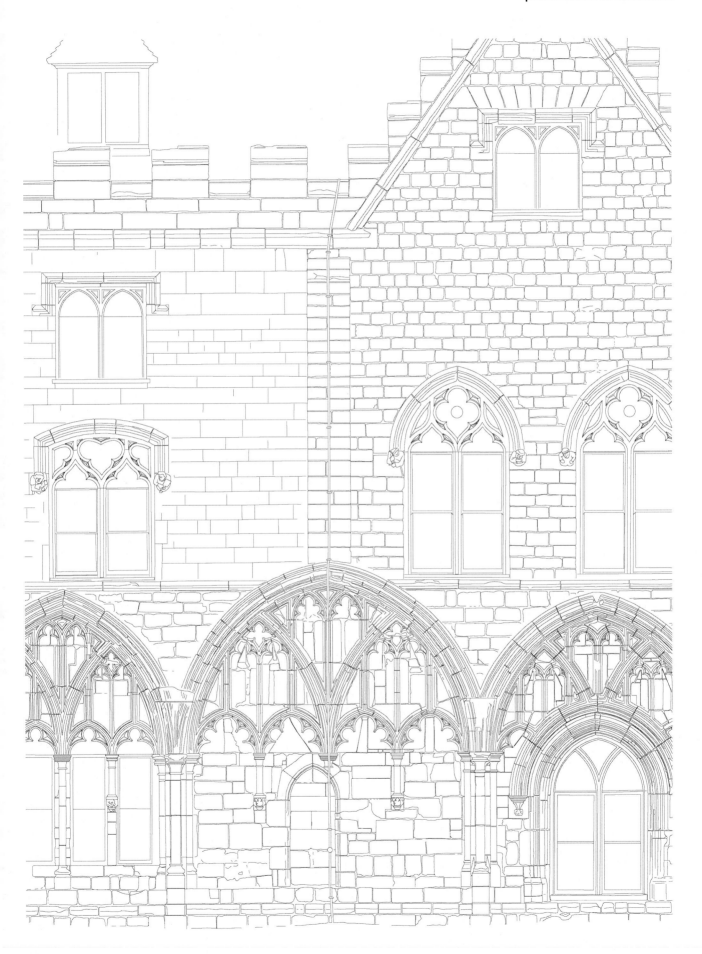

Fig 4.2 Outline detail photogrammetric survey originally prepared for presentation at 1:20 scale

Fig 4.3 Orthophotograph derived from ground-based photography

Fig 4.4 Orthophotograph derived from SUA aerial photography

Fig 4.5 Single-scaled rectified image

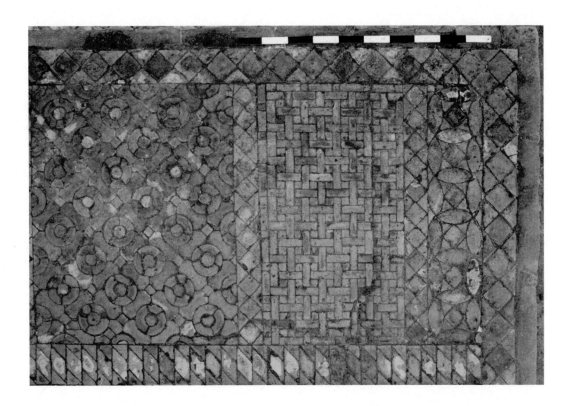

Fig 4.6 Coordinate-controlled rectified image

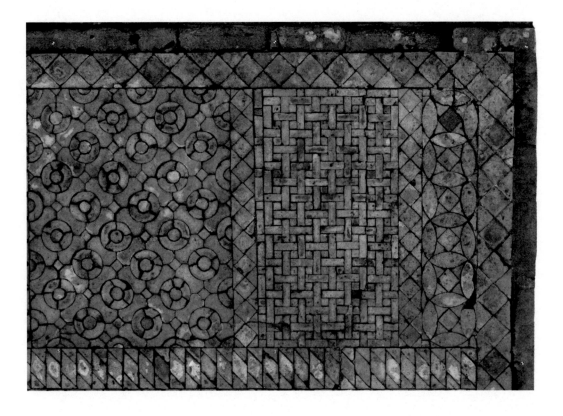

Fig 4.7 Rectified digital montage

Fig 4.8 Building edge detail, showing faced stonework and mortar

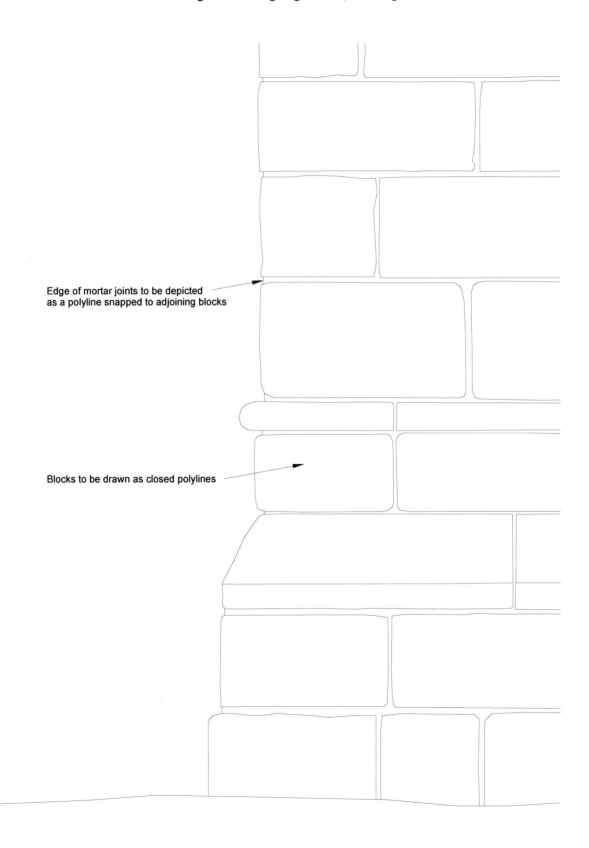

Edge of mortar joints to be depicted
as a polyline snapped to adjoining blocks

Blocks to be drawn as closed polylines

Fig 4.9 Use of layers and line types

0P-Major   0P-Opening   0P-Core

detail obscured
by vegetation

0P-Sculptural   0P-Text   0P-Architectural   0P-Services   0P-Opening

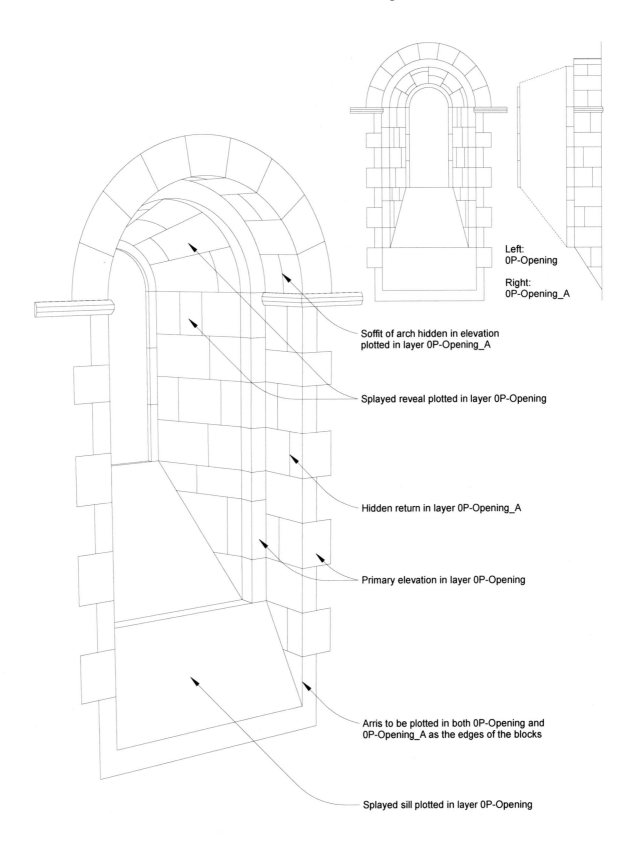

Fig 4.10 Presentation of reveal surfaces

Left:
0P-Opening

Right:
0P-Opening_A

Soffit of arch hidden in elevation
plotted in layer 0P-Opening_A

Splayed reveal plotted in layer 0P-Opening

Hidden return in layer 0P-Opening_A

Primary elevation in layer 0P-Opening

Arris to be plotted in both 0P-Opening and
0P-Opening_A as the edges of the blocks

Splayed sill plotted in layer 0P-Opening

Section 5

Standard specification for measured building survey

5.2 Description of products

The required products must be correctly described so that there is clear agreement about terms such as 'section' and the completeness expected (full height etc), particularly if an indirect technique such as photogrammetry or laser scanning is proposed. Suppliers must not expect the client to accept unfinished work on the basis of constraints of the capture method.

A plan (Figs 5.1 and 5.2) is a convention for showing the horizontal extent of a building. A cut-line is required to show the walls of the building. The convention is for the cut-line to follow the height of a line between hip and shoulder height of a person standing. The cut-line is not simply a height at which the plane of projection is set, for it can vary. Clients are advised to closely specify a desired cut-line if there is ambiguity over the suitable height of the line (eg at changes in floor level or where buildings are built into a slope).

A section line can be taken anywhere through the building (Fig 5.3). The section line defines a plane of projection for the preparation of an elevation view. Section lines must be clearly defined in terms of position, extent and direction of view. They can be adjusted to include or exclude features (eg chimneys), but the line must remain parallel to the original plane. The exact position of the section lines needed to show the required aspects of the building when projected as a sectional elevation should be clearly delineated on sketch diagrams to accompany the project brief (see section 1.1.10).

Sections and sectional elevations are different (Fig 5.7). Determining the cut-line and the direction of view as well as the detail to be included is important. A simple profile can be referred to as a section but is often taken to mean a full-height sectional elevation. It is essential that the terms are used correctly, as there is a great deal of work involved in preparing sectional elevations, which, once started, cannot be easily changed without expense, particularly the view or the position of the section line.

5.1 Measured building survey

5.1.1 Definition of measured building survey

For the purpose of this specification 'measured building survey' is defined as the supply of metric survey data pertaining to buildings and presented as plans, sections, sectional elevations and elevations.

5.2 Description of products

The survey is to be supplied as a CAD drawing in the form of plans, sections, sectional elevations and elevations presented graphically (ie using lines and symbols). Where necessary the graphical data should be supplemented by text annotation (eg description of floor covering and material, height information). The correct use of line type, line weight and layers is essential in order to present the drawing elements in accordance with architectural convention. The building subject is to be presented using an orthogonal projection (ie the plan, section, sectional elevation or elevation is to be shown as a parallel projection onto a horizontal or vertical reference plane as described below).

5.2.1 Plan

A view of the structure as seen in a horizontal reference plane defined by the cut-line. The plan will show information above and below the reference plane unless this information is covered on another plan. The cut-line will reveal full architectural detail, deformation or displacement both at the height of cut and also above and below it. It should be made as informative as possible by cutting across door and window openings.

5.2.2 Section

A view of the internal space of the subject showing only those elements (including the thicknesses of walls) cut by a vertical reference plane.

5.2.3 Sectional elevation

A view of the internal space of the subject as seen from a plane defined by the cut-line or section line and showing all detail revealed by that view. Major structural components not visible (eg hidden from view or in front of the cut-line) may be required to be shown by use of a dashed line.

5.2.4 Elevation

A view of a facade or wall of the subject as an orthographic projection.

5.3.3 Accuracy of survey data

For an explanation of rmse, *see* section 2.1.2. Applying the requirements of this clause to measured building surveys at the standard architectural scales means:

- At 1:50 a 0.3mm rmse is equivalent to 15mm at actual size, therefore 68.3 per cent of points in a representative sample must be accurate to ±15mm. Errors exceeding ±45mm are to be regarded as mistakes.

- At 1:20 a 0.3mm rmse is equivalent to 6mm at actual size, therefore 68.3 per cent of points in a representative sample must be accurate to ±6mm. Errors exceeding ±18mm are to be regarded as mistakes.

5.3 Control for measured building surveys

5.3.1 Control of survey data

The control of measured building surveys is to be achieved principally by use of an adjusted traverse network and must meet the performance described in section 2.2. However, this may be supplemented by the use of control methods suited to graphical survey techniques to achieve the necessary distribution of control points. The use of such techniques must be highlighted in the method statement and included in the survey report.

Appropriate orders of control

The provision of survey stations by a method without full rigorous observation (eg the extension of a ground traverse to upper level floor plans) is, however, unacceptable.

5.3.2 Local datum

Local datum points must be transferred from the site vertical datum by an appropriate method such as surveyors' level or theodolite observation and recorded on the drawing as:

- a vertical datum plotted at metre intervals; or

- a reference datum line marked with the datum value; and/or

- annotation of detail with recorded height.

Heights on floor plans

Where plans for more than one floor level are required the heights shown for each floor must be given relative to a single datum. Multiple arbitrary datum points for each floor must not be used.

5.3.3 Accuracy of survey data

The plan position of any well-defined detail shall be accurate to ±0.3mm rmse at the specified plan scale when checked from the nearest survey control station.

To verify the achievement of the specified tolerances, the following may be required:

- booked data showing directly measured dimensions;

- coordinate data and their provenance, where dimensions between points have not been directly measured.

5.3.4 Precision of detail measurement

The precision of detail measurement is to be as specified in section 2.1.2.

5.4 Drawing content

5.4.1 Detail required

The required scale of survey will determine both the level of detail and the expected precision. The level of detail refers to the density of information, while precision refers to the performance of the measured points used to delineate the detail. At a larger scale, such as 1:20, a plan, section or elevation will show more information than at a smaller scale, for example 1:50 or 1:100.

Detail comprises the visible features delineated within a plan, section or elevation such as openings, straight joints, roof scars, the jointing of masonry, the outline of fittings and fixtures or the outline of materials used. Sectional detail is to include eaves, sills, lintels, sashes etc.

1:100 and 1:50 scale

The smallest plottable detail is 0.2mm (at 1:50 scale this equates to a 10mm × 10mm object), so a degree of generalisation is required.

- Large linear objects, such as skirting or cornices, must be shown as a light line inside the wall or cut line.

- Annotation indicating floor material and direction of floorboards is to be included. A single line can be used to show joints in timber or for floor coverings.

- Openings in plan may be generalised, but must show an indication of the type of detail by careful use of an approved symbol for sash, mullion, door swing and lining. For elevations at 1:100, repetition of a single measured window is permitted in cases where they are demonstrably similar.

- Overhead detail, such as beams, vaults, stair flights, reveals etc, must be shown as a dashed line.

1:20 scale

All detail and annotations that would appear at 1:50 will also be present at 1:20. In addition, all visible architectural features must be shown, including:

- mouldings and sculptural detail from actual size source material (such as a profile trace or measured drawing);

- all stone by stone detail and galleting for elevations;

- floor detail such as the plan of stone flags or floor tiles for plans;

- timber components with pegs, peg holes and open or re-used joints plotted using a separate line to describe each component;

- eroded edges as seen in the required view to show the condition of the fabric;

- the deformation of wall surfaces at the cut-line and foot of wall line; and

- openings in full detail as apparent from the plane of reference.

5.5.1 #Curved features

Choose an option. A curved or conical facade may be projected as unwrapped to show the entire facade true to scale on the plot. If the facade is required to be seen as 'square on', the edges will be foreshortened.

5.5.2 # Depiction of cut-line (plan and section)

The position and direction of view of the section line must always be shown on a plan or key plan – choose an option.

Building footprint

The building footprint or ground or floor line is the line at the foot of the wall. Plans of vertical walls that have a constant width over their full height will not show this line unless it is specifically requested. Where a wall has a batter or sits on a plinth, the line will be visible and should appear on the plan.

Choose an option for depiction of the building footprint.

5.5 Drawing convention

5.5.1 Curved features

Curved features should be presented either:

(a) unwrapped so as to provide a true-to-scale representation; or

(b) as an orthogonal view.

The method proposed for any required unwrapping of data must be outlined in the method statement.

5.5.2 Depiction of cut-line (plan and section)

The cut-line(s) must be shown with a line weight of a thickness determined by the output or plot scale.

Sections and sectional elevations

The cut-lines of any sections or sectional elevations should be clearly shown on either:

(a) the accompanying plan; or

(b) a key plan.

The line must include arrows showing the direction(s) of view (Fig 5.3).

Building footprint

The contact lines between the building and the ground (also known as the ground line, when visible in elevation) must to be shown with a lighter line than the cut-line. The visibility of the line will depend on the wall, its inclination and the required scale.

The building footprint is to be:

(a) shown; or

(b) omitted; or

(c) recorded in 3-D in a frozen layer.

5.5.3 #Use of symbols

Alternatives may be agreed as required.

5.5.3 Use of symbols

Symbols may be used as tabulated below. Level and dimension values are to be shown to two decimal places throughout.

item	scale	size on plot	symbol
door swing	1:20	full extent of swing	shown as an arc
	1:50	open at 90° or 45°	
levels		2mm cross, text 2mm plot height	cross with value to top right
step direction		text 2mm plot height	arrow pointing up direction of run, labelled 'up'
glazing detail		0.25mm line	single line on centre of window frame; frame beads omitted
room height	1:20 and 1:50	text 2mm plot height	enclosed in an ellipse
window/door opening height		text 2mm plot height	small upward and downward pointing open arrow heads
window/door soffit/lintel height		text 2mm plot height	small upward pointing open arrow head
window/door sill/ threshold height		text 2mm plot height	small downward pointing open arrow head
roof survey – direction of fall			arrow pointing down slope
windows and doors	1:100	repetition of a single measured type permitted	

5.5.7 #Use of text

Choose the preferred font.

5.5.4 Point density and line quality

Point density and line quality is to be in accordance with the performance specified in Section 2.

5.5.5 Use of 'best profile'

The depiction of architectural forms requires special attention to the detail of functional openings such as sills, door openings, splays, mullions, plinths etc. Mouldings must be shown as completely as possible, with the 'best profile' shown. Where a profile of a damaged or eroded moulding can be derived with certainty it should be shown 'as complete' with the cut-line profile shown as a dashed line.

5.5.6 Assumed detail

Assumed detail should be presented using dashed lines, clearly indicated and on a separate layer. If detail is absent from a drawing, then the space is to be annotated with an explanation (eg 'no access', 'obscured at time of survey' etc).

5.5.7 Use of text

Text is only to be used if the information needed cannot be displayed as a graphic component of the drawing. Use of text is restricted to:

- annotation of direction of steps;

- description of material and services using appropriate abbreviations;

- values of spot heights, room heights etc;

- notification of restrictions to survey (see section 5.5.6);

- as required by Section 3.

The text height is to be 2mm at the plot size.

The text style is to be either:

    (a) Arial; or

    (b) other (specify).

Text is to be positioned on the drawing such that it is:

- aligned with the sheet edge if possible;

- aligned with large linear objects;

- as close as possible to the object described;

- not overlapping or breaking plotted lines; and

- preferably to the upper right of the object described.

If the upper right default position causes text to be in conflict with detail or other text, it is to be placed elsewhere in the following order of preference:

    1 upper left

    2 lower left

    3 lower right

    4 rotated at default position to avoid clash

5.5.8 #Overhead detail on plans

Choose an option. It is recommended that overhead detail is included, as its omission will limit the usefulness of the final product.

5.5.8 Overhead detail on plans

Large-scale surveys will require the depiction of the principal features of overhead structures such as vaults, beams, gantries, ceiling details, high level windows, roof lights, pulleys, murder holes etc. The annotation 'at high level' or '(at HL)' can be used to indicate detail above the plan height if it is not clear from the plotted lines alone.

Vaults, at 1:50 and 1:20 scale, should be shown by a plot of the rib lines, with imposts and bosses in outline. A single dashed line indicating the centre line of the rib may be used at 1:100 scale.

Overhead detail is to be:

    (a) recorded in 3-D and plotted at true height; or

    (b) plotted in 2-D congruent with all other plan detail; or

    (c) omitted (not recommended).

5.5.10 #Treatment of staircases on plans

Staircases must be shown. The amount of detail will vary according to the required scale. All staircases will require the use of a break line to show the intersection of the stair with the cut-line for the plan (Fig 5.5). Indicate the required options for levels and annotation.

5.5.11 #Services

Choose the required details or add/delete from the list as necessary.

Historic England

5.5.9 Floor detail on plans

Plans at 1:20 and 1:50 scale are required to show the following floor details:

- changes in floor treatment;

- changes in floor level;

- steps: the line of tread noses (continuous) and risers (dashed, if undercut); and

- flagstones etc, depending on scale.

Fixings to walls and floor as seen on the cut-line (hinges, sockets, niches etc) should be shown in a line thickness greater than that used to depict all other detail.

5.5.10 Treatment of staircases on plans

The required convention for the depiction of stairs is to show the plan as seen from the cut-line and to use a break line to show the interruption of the plan, (Fig 5.5). Where stairs include detail such as half landings between floors that would not otherwise appear on a drawing, an inset plan is to be used. Overhead detail is to be shown as required by section 5.5.8.

Levels on steps and stairs should be shown either:

- (a) on each landing (ie at the top and bottom of each flight); or

- (b) on all treads.

Stairs are:

- (a) to be annotated with numbers to each tread; and/or

- (b) annotated with 'up' arrow as described above (5.5.3); or

- (c) not to be annotated.

5.5.11 Services

Large components such as radiators, exposed pipe-work, shafts, ducts etc must be shown in full detail. Smaller components may be indicated by standard symbol and/or annotation. The following services details must be shown and annotated with service type:

- large fittings only;

- pipe-work;

- rainwater goods;

- duct-work;

- electrical fittings (in elevations only).

Electrical wiring and fittings are not usually required to be shown on plans unless specified in the brief.

5.5.12 #Levels

Choose a preferred method for the indication of door and window heights.

5.5.13 #Roof survey

Choose an option for any roof survey (Fig 5.6).

Historic England

5.5.12 Levels

Levels must be shown relative to the vertical datum as specified in Section 2.

Levels must be located at the following locations where applicable:

- thresholds;

- either side of door openings;

- centre of each room;

- in each corner of each room;

- interior sills;

- exterior sills on centre of sill boards; and

- lintel soffit.

The heights of window and door openings shall be either:

    (a) as indicated by soffit/lintel and sill heights; or

    (b) shown as an opening height.

Floor to ceiling heights are required for each room and are to be shown enclosed in an ellipse.

5.5.13 Roof survey

Roof survey drawings can be presented in one of two states: either with the roof cover (slates, tiles, lead etc) on or with the roof cover off. A survey may be required to show rafters and trusses or trusses only.

A roof plan is required showing:

    (a) 'cover on'; or

    (b) 'cover off'; or

    (c) 'cover off trusses only'.

In all cases the roof must be shown in plan, ie looking straight down.

Appendix 5.1 CAD layer names for measured building survey

Appendix 5.1 CAD layer names for measured building survey

layer	line type	line weight (mm)			description
		1:20	1:50	1:100 or above	
0A-anno	continuous	0.13	0.13	0.13	all annotation not associated with levels, grid, services, drawing sheet format
0A-chimney	continuous	0.35	0.25	0.25	
0A-cutline	continuous	0.5	0.35	0.35	the cut or plan line
0A-detail	continuous	0.13	0.13	0.13	lines used to plot detail; if detail overlaps an edge, only the heavier line weight is to be used
	dashed	0.13	0.13	0.13	below plan detail, inside cutline
	continuous	0.18	0.13	0.13	below plan detail, outside cutline
	dashed	0.25	0.18	0.13	projected – in front of section line
	dashed	0.13	0.13	0.13	within cutline of section, eg niche, window etc
0A-digi	continuous	0.13	0.13	0.13	digitised from other sources
0A-grid	continuous	0.13	0.13	0.13	grid points to be frozen on presentation
0A-gridtxt	continuous	0.13	0.13	0.13	all text associated with the grid: annotation to be aligned with the grid line
0A-hidden	dashed	0.13	0.13	0.13	2mm line, 1mm spacing
0A-inst_cntl	continuous	0.13	0.13	0.13	control positions – to be frozen on presentation
0A-joist	continuous	0.13	0.13	0.13	
0A-level	continuous	0.13	0.13	0.13	
0A-level_text	continuous	0.13	0.13	0.13	
0A-trav	continuous	0.13	0.13	0.13	control diagrams – to be frozen on presentation
0A-overhead	dashed	0.13	0.13	0.13	overhead detail – 2mm line, 1mm spacing
0A-opening	continuous	0.25	0.18	0.18	lines describing the edges of openings, changes of plane or skyline
0A-plinth	continuous	0.13	0.13	0.13	plinth lines in plan, inside cutline
	continuous	0.25	0.18	0.13	plinth lines in plan, outside cutline
0A-purlin	continuous	0.13	0.13	0.13	
0A-rafter	continuous	0.13	0.13	0.13	
0A-svs_elec	continuous	0.13	0.13	0.13	
0A-svs_fire	continuous	0.13	0.13	0.13	
0A-svs_foul	continuous	0.13	0.13	0.13	
0A-svs_gas	continuous	0.13	0.13	0.13	
0A-svs_other	continuous	0.13	0.13	0.13	
0A-svs_water	continuous	0.13	0.13	0.13	
0A-text	continuous	0.13	0.13	0.13	

0A-title	continuous	0.35	0.35	0.35	
0A-truss	continuous	0.25	0.25	0.25	
0A-wallplate	continuous	0.13	0.13	0.13	
Additional layers for vault surveys					
0A-boss	dashed	0.13	0.13	0.13	bosses may be shown as an outline
0A-cap	dashed	0.13	0.13	0.13	capital, impost or abacus
0A-corbel	dashed	0.25	0.25	0.25	on plans, if at high level, shown as an outline
0A-rib	dashed †	0.13	0.13	0.13	for the rib lines; to be expanded according to rib type if required
0A-shaft	dashed	0.5	0.35	0.35	on plans, usually shown as a cut-line; on sections, use detail line weight

† dashed on plans, continuous for elevations

This is not an exhaustive list. New layers may be created so long as they are prefixed with 0A-.

The cut line of a building or feature should be of a heavier weight than lines used for other detail.

A dot and peck line type should be used to indicate any or all of the line types in the table if there is a conflict of lines and for boundaries if required to avoid confusion. The dot and peck line should comprise a line 1mm in length separated by a 0.5mm gap from a dot of 0.18mm, with a 0.18mm line width.

A dotted line may be used for clarity if there are a large number of dashed lines on the drawing sheet.

In AutoCAD, LTGEN is to be set to on.

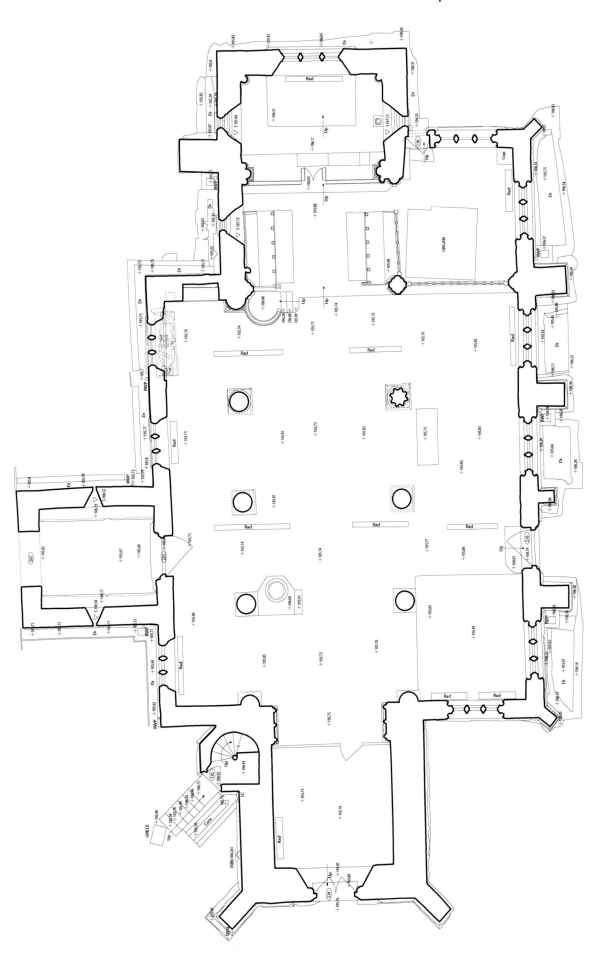

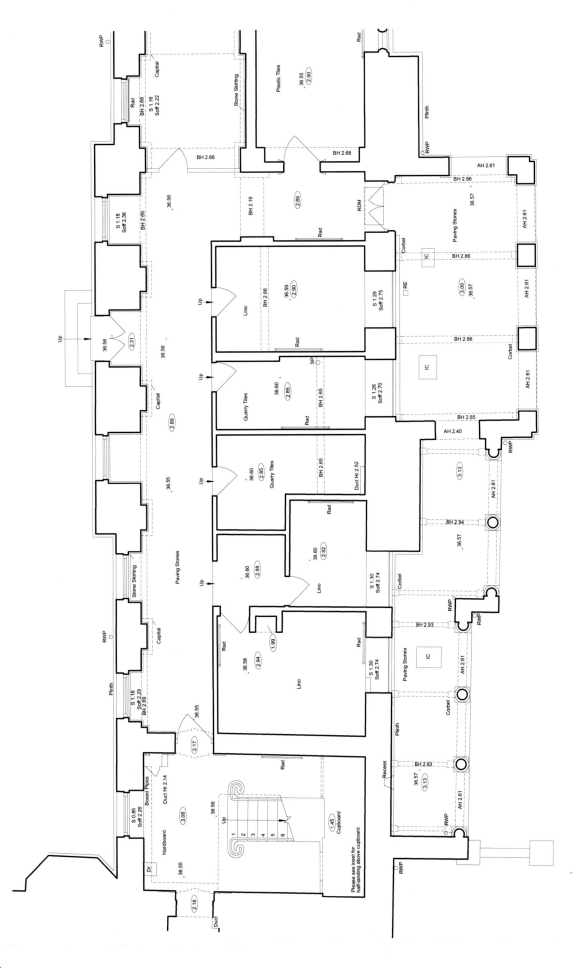

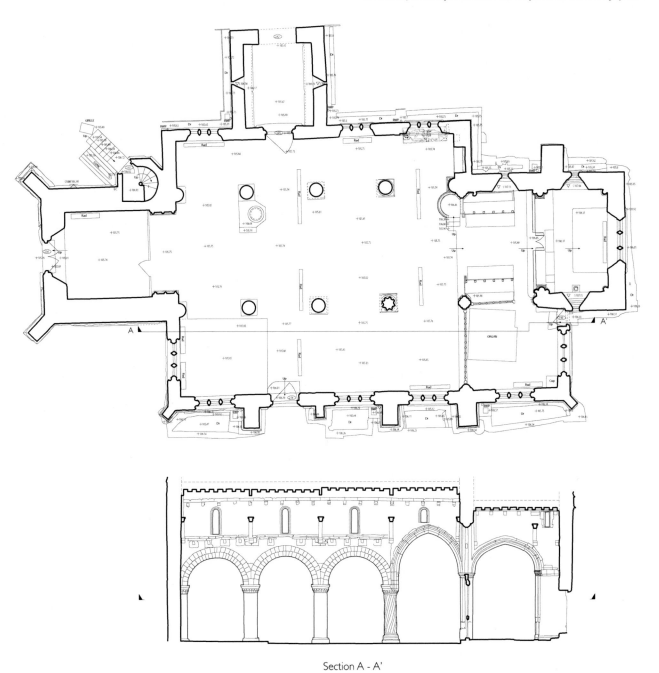

Fig 5.3 Section lines; top, shown related to plan; bottom, multiple sections depicted on a key plan

Section A - A'

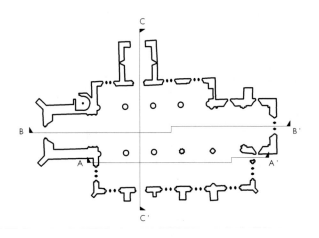

Fig 5.4 Examples of overhead detail; top, plan with a vaulted ceiling – 1:20 level of detail; bottom, overhead beams on a plan – 1:50 level of detail

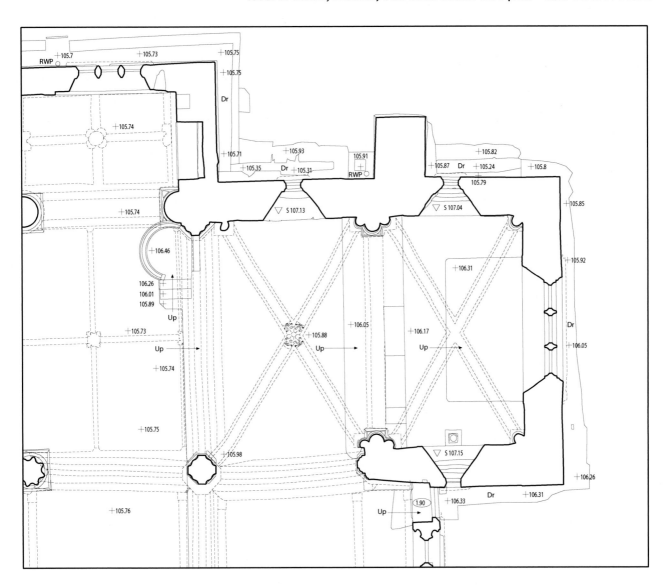

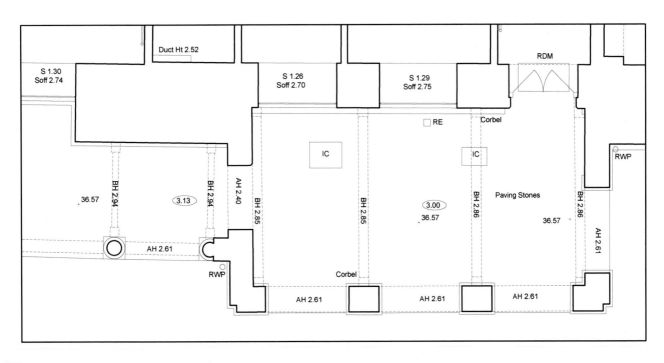

Fig 5.5 Examples of the treatment of staircases on plans

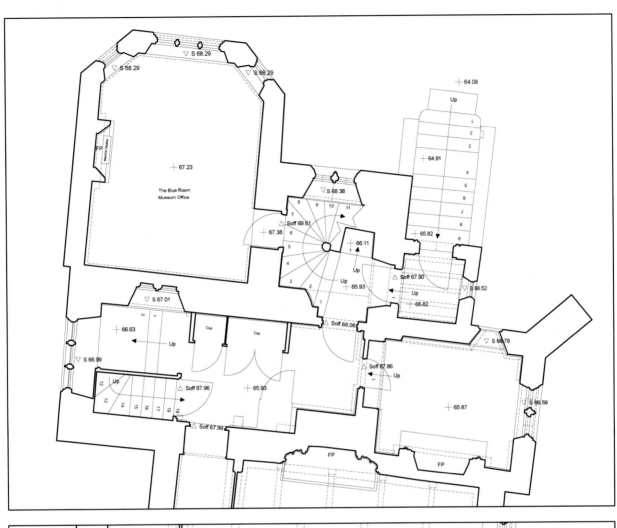

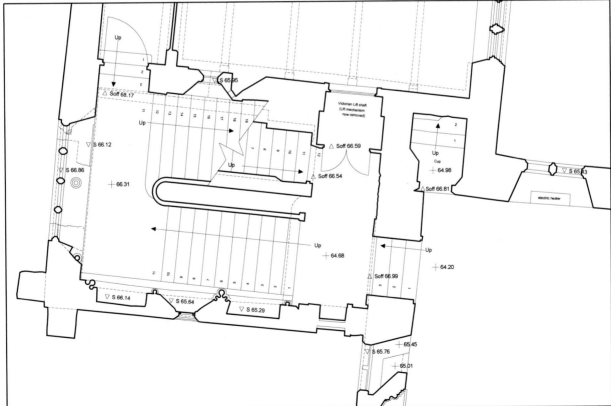

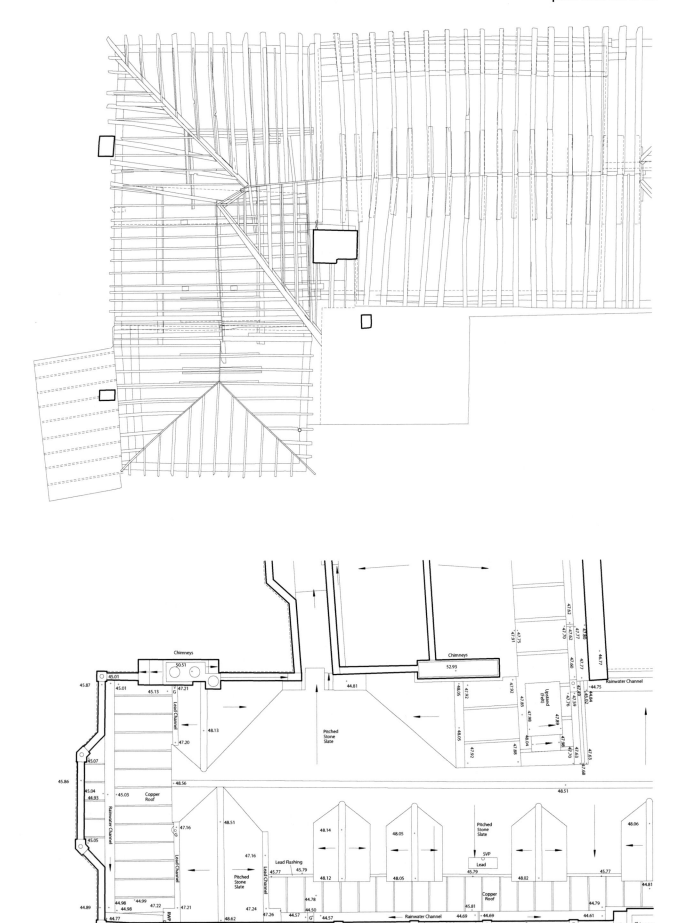

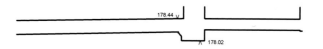

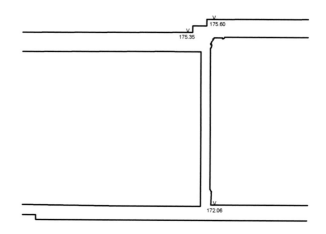

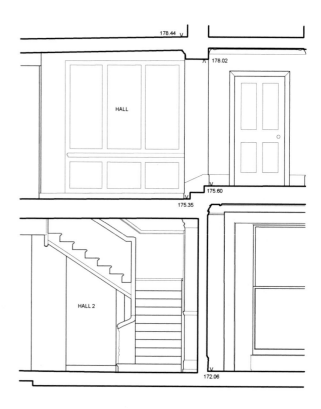

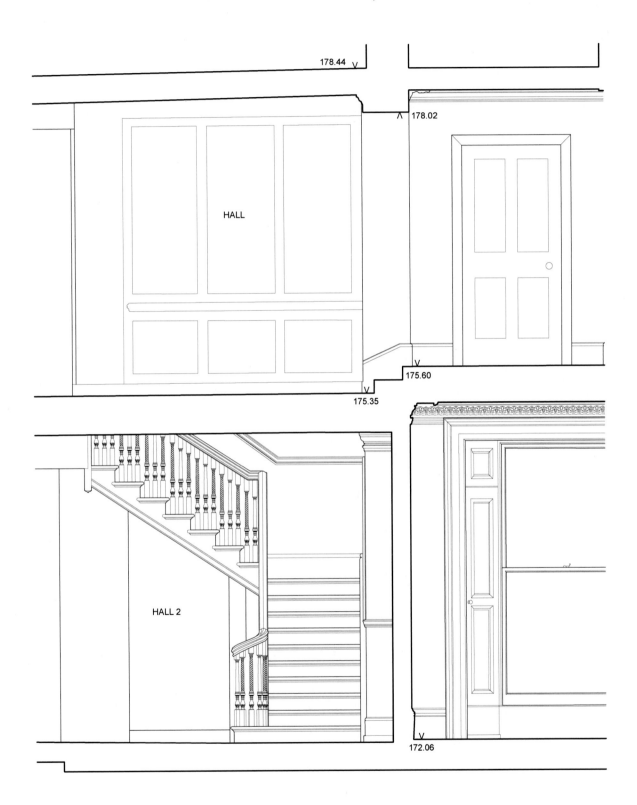

Fig 5.7 Section and sectional elevations: top left, profile or section;
bottom left, sectional elevation – 1:50 level of detail;
below, sectional elevation – 1:20 level of detail

178.44 ∨

∧ 178.02

HALL

∨
175.60

∨
175.35

HALL 2

∨
172.06

Section 6

Standard specification for topographic survey

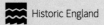

6.1.1 #Definition of topographic survey

Choose an option. A 3-D survey will be more expensive but may prove much more useful than a 2-D product.

6.2 #Description of products

Choose options or edit as appropriate

Plan

For extensive surveys, a projection and scale factor may be applied to ensure congruence with mapping from other (eg the Ordnance Survey) sources (*see* section 2.2.4). These must be described in the survey documentation. For vertical orthographic projections, scale factor will be 1.

Profile

A vertical exaggeration of greater than one will emphasise the nature of the terrain but should be used with caution as the height values will no longer be to scale.

Detail

That is, everything that is not control, contours, spot heights or hachures.

Both hard and soft detail should be annotated with heights to two decimal places.

6.1 Topographic survey

6.1.1 Definition of topographic survey

For the purposes of this document topographic survey is defined as the controlled measurement of natural and artificial landscape features. It is to be presented as either:

(a) a two-dimensional (2-D); or

(b) a three-dimensional (3-D)

dataset reading as a plan. Profiles and a digital terrain model (DTM) may also be required (*see* section 6.6).

6.2 Description of products

For the purpose of producing large scale topographic survey, the following definitions apply.

Plan

This will be either:

(a) a vertical orthographic projection onto a horizontal reference plane; or

(b) a cartographic projection and a scale factor may be applied.

The plan may incorporate information above and below the reference plane; buildings shown will normally be cut on a different horizontal plane to that used for the main plan. The view is to be presented both as plotted or drawn 'hard copy' such that there is no discrepancy beyond permitted standard error (*see* section 2) and as a CAD file containing the same data as the plot.

Profile

A horizontal orthographic projection onto a vertical reference plane. Profiles will show the surface or cross section of the ground, including the thickness of any walls. The end points and line taken by the profile(s) should be clearly marked on a key plan. The vertical exaggeration for such views shall be ×1 unless otherwise stated.

Detail

The visible features, excluding the surface of the terrain, shown on the plan. It may be considered either 'hard' or 'soft'.

Hard detail is that defined with a clearly visible edge eg a kerb.

Soft detail has an undefined edge or surface eg earthworks.

6.3.2 #Adequate site cover

Choose (b) if stations are required in certain areas to allow, for example, later re-occupation to facilitate a subsequent survey. Include a diagram showing where stations are to be located. It may also be necessary to show areas where stations are not to be inserted.

6.3.3 Contours

See section 2.1.2 for an explanation of rmse.

If, for example, the contour interval is to be 0.25m, then 68.3 per cent of a representative sample of points forming a contour should be correct to better than ±0.08m and 95.4 per cent to better than ±0.13m.

6.3.4 Spot heights

If the contour interval is to be 0.25m, then 68.3 per cent of a representative sample of spot heights should be correct to better than ±0.06m and 95.4 per cent to better than ±0.1m.

6.3 Survey control

6.3.1 Coordinate system

Where use of the OSNG is specified, the primary site control, or starting coordinates for it, is to be established by means of GNSS observation. The WGS84 values are to be transformed to the OSNG using the OSTN02 transformation. Height values are to be transformed using the OSGM02 transformation. The scale factor used must be reported in the survey documentation.

The method statement must describe the equipment and procedures to be employed so as to achieve the precision specified in section 2.1.2.

Secondary control may be achieved by traverse observations.

6.3.2 Adequate site cover

The control network or traverse must extend so that stations are in reasonable proximity to the perimeter of the survey area and the detail to be mapped.

The distribution of stations is to be either:

(a) at the discretion of the surveyor; or

(b) decided in consultation with the client.

6.3.3 Contours

Contours shall be correct to an rmse of better than one third of the contour interval, where a representative sample of points on contour lines is checked by precise measurement from the nearest control point (and hence 95.4 per cent of a representative sample shall be correct to better than half of the specified contour interval).

6.3.4 Spot heights

Spot heights shall be correct to an rmse of better than one quarter of the contour interval, where a representative sample is checked by precise measurement from the nearest control point (and hence 95.4 per cent of a representative sample shall be correct to better than 0.4 of the specified contour interval).

6.3.5 Precision of detail measurement

The precision of detail measurement is to be as specified in section 2.1.2.

6.4.1 #Scale

Choose the required scale(s)

'Overscale' survey

For example, where a group of buildings or trees require an enhanced level of definition on the plan, at a scale of eg 1:100.

Historic England

6.4 Detail required

6.4.1 Scale

Topographic survey is required at a scale of either:

    (a) 1:200; and/or

    (b) 1:500; or

    (c) as specified in the project brief (section 1.1.10).

If the survey is to be presented at more than one scale, different sizes of text are to be layered in the CAD file so that they may be segregated to allow for legibility at both scales.

It may be necessary to carry out some of the survey at a larger scale than that commensurate with the plot scale. Reference will be made in the project brief as to the area and nature of 'overscale' survey required. At 1:500 scale, a degree of generalisation from the 1:200 level is acceptable. The smallest plottable detail is 0.2mm × 0.2mm, which equates to 100mm × 100mm at 1:500 scale, therefore symbols should be used to describe visible features smaller than this.

For all hard detail the accuracy of planimetric information shall be such that the plan position of any point shall be correct to within +/-20mm rmse when checked from the nearest permanent control station when surveyed for presentation at scales between 1:100 and 1:200. At 1:500 scale, any point of hard detail shall be correct to within +/-30mm rmse.

6.4.2 #Detail required

Edit as required.

6.4.3 #Obstructed ground

Select the option required. Detail or contours that cannot be surveyed without clearing will be treated accordingly. In many cases it will be more economical for the client to arrange clearance.

6.4.2 Detail required

The following general categories of information shall be surveyed:

- roofed buildings/structures (Fig 6.1);

- roofless/ruined structures;

- temporary/mobile buildings;

- visible boundary features: walls, fences and hedges;

- roads, trackways, footways and paths;

- street furniture;

- statutory authorities' plant and service covers where visible;

- changes of surface;

- isolated trees/wooded areas/limits of vegetation (Figs 6.2 and 6.3);

- pitches/recreation;

- private gardens or grounds (off-site areas);

- water features;

- earthworks;

- industrial features;

- railway features (with arranged access);

- above ground services;

- underground services (Fig 6.5); and

- other (specify).

All of the above are to be presented using the specified cartographic conventions (*see* section 6.6), and either drawn or depicted using symbols dependant on output scale. The plans should have a fixed control network and measurable repeatability of precision commensurate with the required scale (section 2.1.2).

6.4.3 Obstructed ground

Details or contours that cannot be represented to the specified accuracy without extensive clearing shall be:

    (a) surveyed approximately and annotated accordingly; or

    (b) surveyed, following clearance by the client; or

    (c) surveyed, following clearance by the survey contractor.

6.5 Underground services

(A) Record information

Existing information taken from record plans covering underground services is often incomplete and of doubtful accuracy. It should usually be regarded only as an indication and cannot be guaranteed.

(B, C) Underground services surveys

Drainage covers should not be lifted without the permission of the owning authority. Many local authorities do not permit their inspection covers to be lifted but will provide some information for a standard fee.

(D) Electronic tracing

This is a more reliable method of locating buried services. On heavily built-up sites 85 per cent completeness is probably all that can be expected. Plan accuracies of +/-150mm may be achieved but this will be dependent on the depth of the service below ground level. Where similar services run in close proximity, separation may be impossible. Successful tracing of non-metallic pipes may be limited.

Further guidance, if required, on utility surveys can be obtained free of charge from The Survey Association (TSA) at:

> **www.tsa-uk.org.uk/for-clients/guidance-notes/**

For a comprehensive specification, *see:*

PAS 128:2104 Specification for underground utility detection, verification and location, British Standards Institution 2014, ISBN 978 0 580 79824 5

available from the BSI shop **http://shop.bsigroup.com/**.

6.5.1 #Extent of survey required

Tick in the table the type of survey required.

6.5.2 #Services information

Where Section 6.4.1 includes survey by (B), (C) or (D), select the required method for recording the information.

6.5 Underground services

An accurate base plan is essential for the plotting of underground utility services. If such a plan does not exist it will be necessary to produce one (Fig 6.5).

Underground services surveys will be undertaken using one or more of the following methods:

(A) Consulting underground service records. (To be taken from statutory or other authorities' record drawings and plotted to agree as closely as possible with surveyed surface features.)

(B) Direct visual inspection. (Accessible inspection chamber covers should be lifted where permissible and services positively identified.)

(C) Direct visual inspection supplemented by consulting service record drawings. (Accessible inspection chamber covers should be lifted where permissible and services positively identified. Routes of services between access points to be taken from record drawings and plotted to agree as closely as possible with surveyed surface features and trench scars where obvious.)

(D) Full investigation, including electronic tracing. (Services to be fully investigated by visual survey supplemented by electronic or other tracing of inaccessible routes.)

6.5.1 Extent of survey required

Services listed below shall be surveyed by the method indicated (*see* above for description of methods). All work should be carried out with due regard to the Health and Safety guidelines for working within confined spaces.

A	B	C	D	service
				surface water drainage
				foul drainage
				water
				gas
				electricity
				telecommunications
				other services
				other underground features (specify)

6.5.2 Services information

Information derived from survey methods (B), (C) and (D) shall be supplied as either:

(a) invert levels, pipe diameters and annotations on drawings or digital files; or

(b) inspection chamber description sheets.

The date of inspection/survey must be included.

6.5.3 Derived information

Where information is derived from statutory authorities' record drawings, a schedule shall be provided giving full details (eg drawing number, scale, date etc). All information taken from records shall be clearly identified as such in the survey product and placed on a separate layer.

6.5.4 Report

A report shall be submitted indicating any anomalies between surveyed data and records, detailing likely accuracies achieved and commenting on services not located for any reason (eg unliftable or hidden covers). All identified features should be highlighted in this report.

6.6.1 #Landform, earthworks and surface terrain

Delete any that are not required.

6.6.2 #Contouring and DTM

- *Plans up to 1:200:* Select the option required.

- *Plans at 1:500:* Select the option required.

- *Index contours:* Indicate the index contour frequency required.

6.6 Drawing convention

6.6.1 Landform, earthworks and surface terrain

Landform, earthworks or surface terrain are to be indicated by:

- surveyed contour;

- form line;

- annotation;

- spot height; and

- hachure.

6.6.2 Contouring and DTMs

Contours are required to represent the surface characteristics of the terrain. They are to be shown with contour values reading up the slope at a density sufficient to identify all contours without ambiguity. Where contour values are inserted the contour lines must be broken to ensure legibility. The contours must be shown cut by buildings and structures, including the batter of masonry fortifications built into earthworks. Contour lines must be appropriately smoothed after interpolation to avoid lines with sharp changes in direction (Fig 6.6).

Contours derived from a DTM must not reveal the geometric model used to construct the surface. Care must be taken to ensure that the presence of detectable edges is only a result of such edges being part of the landscape. Breaklines shall be used to ensure that the DTM accurately describes the landform to be depicted by identifying changes of slope at, for example, the tops and bottoms of ditches and banks. When earthworks are mapped, attention must be paid to the surface and its intersection with objects such as gun emplacements, battered walls, chimneys etc, so that a plan of the building components can be seen clear of the contours used to describe the earthworks or landform surrounding them. For the required accuracy of contours (see section 6.6.3).

Plans at 1:100 and 1:200 scale are to be contoured at a vertical interval of either:

    (a) 0.25m; or

    (b) other (specify).

Plans at 1:500 scale are to be contoured at a vertical interval of either:

    (a) 0.5m; or

    (b) other (specify).

Thicker index contours are to be shown at multiples of ......m.

Hachures may be used to supplement contoured information and to describe sub-contour detail (Fig 6.4).

Sufficient levels for the DTM shall be surveyed such that the ground configuration, including all discontinuities, is represented on the survey plan.

The maximum spacings for DTM points are:

scale	ground spacing	distance on plan
1:100	5m	50mm
1:200	10m	50mm
1:500	10m	20mm

Where a DTM is the final product, the density of levels shall be such that the surface of the model is constructed within 0.1m of the true surface when verified by check measurement. The density of levels shall be at least 1m for surfaces with earthworks or 5m for open ground.

6.6.3 #Location of spot heights

In flat areas spot heights will be located at 6–20m intervals for 1:200 scale and at 15–50m intervals for 1:500 scale.

Point descriptor

Select the option required.

6.6.3 Location of spot heights

Spot heights shall be shown in the following positions, except where the ground is obscured by vegetation or other obstructions:

- at salient positions such as top, bottom and along the centreline and mid-point of slopes, ditches, embankments and earthworks;

- at the top and bottom of features described by hachure to support the form lines;

- at significant changes of gradient, along the centre and edges of road, tracks and water courses, at between 50mm and 100mm at map scale;

- in flat areas (where the horizontal distance between contours generally exceeds 30mm at map scale) at intervals between 30mm and 100mm at map scale;

- at the sill tops and thresholds of buildings, ruins and building fragments;

- at the base of walls showing height of ground at the corners, buttresses and change of direction of walls; to include corresponding positions either side of a free-standing wall;

- wall tops on ruined walls, to indicate major changes in wall height and maximum height; large and irregular ruined walls may not require levels other than a general indication of height;

- at regular intervals along dwarf walls, showing the height of ground at the wall base and wall top;

- at changes of surface treatment (eg the edges of grassed areas and hard standing, paths, walkways etc);

- at the surface of drainage inspection covers, the invert level of drainage pipes, on the edge of rain water gullies and along rainwater channels;

- at the edges and high points of large fragments of buildings (fragments of 1m × 1m size or greater on any edge at actual size);

- at the top and bottom (and if practicable on each tread) of flights of steps; and

- at the base of the bole of large trees.

The required control and precision of vertical data is described in Section 2 and at 6.3.

The standard point descriptor must be either:

(a) a cross; or

(b) other (specify).

of no more than 2mm × 2mm at plot size, the intersection of which shall represent the given coordinate value. The symbol is to be aligned with the sheet edge. The point descriptor shall be used for the depiction, with appropriate annotation, of spot heights and reference points. Spot height text shall be 2mm high at plot scale and given to 2 decimal places.

6.6.4 #Depiction of trees and vegetation

CAD layering

See Appendix 6.1 for CAD layering convention.

General points – bole

The base of the bole at ground level is to be shown if there is a significant lean from the vertical.

General points – canopy

The CAD layer containing the trimmed envelope is to be the default visible (plotted) layer, with the layer containing the individual spreads available but not visible in the CAD drawing files.

Tree annotation

Select the options required.

Tree number

Usually located on a metal tag fixed to the trunk at approximately head height.

6.6.4 Depiction of trees and vegetation

Vegetation is to be indicated by a standard scaled symbol and text description of species by common name.

Trees are to be plotted as up to four components: the base, bole, canopy/spread and envelope. Trees are considered to be identifiable as such if they are 5m or greater in height, unless of a species known as a shrub (such as laurel) and lacking an identifiable bole.

If less than 5m, high trees should be depicted as vegetation. Trees, including the bole, are to be shown to scale. Any displacement of the tree canopy from the bole should be shown. Single small trees in unobstructed terrain should be shown even if they may not be of a size that normally qualifies for depiction. Small trees of less than 5m in height are to be layered in the CAD file separately to aid landscape management.

General points

The bole is to be plotted at 1.5m above ground level and to include multiple grouped boles. They are to be a scaled and hatched shape that appears solid on the plotted drawing sheet. The hatching used must be consistent for all bole sizes.

The spread of the canopy is to be shown as a standard scaled symbol. At 1:200 scale and greater, the canopies are to be contained within the digital file such that both the individual spread per tree is shown in one CAD layer and the envelope of a group of trees is shown in another (*see* Appendix 6.1 and Fig 6.2). At 1:500 scale, depiction of the envelope only is sufficient. Where a small tree has an extensive canopy spread over other vegetation, the canopy should be mapped as a dashed line.

Trees are to be annotated with the following information:

- the girth at breast height;

- the tree number where visible;

- species by common name; and

- height to the nearest 0.5m.

For vegetation, hedges are to be depicted using a linear symbol. They shall be surveyed so that the centre line, width and descriptive annotation are clearly shown on the plan (Fig 6.3).

The extent and type of other vegetation is to be shown, annotated in a similar manner to that used for hedges.

6.6.5 #Text style and positioning

Major objects

For example buildings, bastions, named areas etc.

Select the font required.

Historic England

6.6.5 Text style and positioning

For annotation, levels, index contours and descriptions of form or surface treatment the height of text should not exceed 2mm at plot scale. For major objects the text shall be 5mm in height at plot scale.

The font used is to be either:

(a) Arial; or

(b) other (specify).

For drawing sheet title text *see* section 3.3.1.

Text is to be positioned on the drawing such that it is:

- aligned with the sheet edge if possible;

- aligned with large linear objects;

- as close as possible to the object described;

- not overlapping or breaking plotted lines; and

- preferably to the upper right of the object described.

If the upper right default position causes text to be in conflict with detail or other text, it is to be placed elsewhere in the following order of preference:

1 upper left

2 lower left

3 lower right

4 rotated at default position to avoid clash

6.6.7 #Depiction of buildings and walls

Cut line

Select the option required.

Scales of 1:100 or larger – floor detail

For example the plan of stone flags or floor tiles.

Additional detail

Select the option required.

6.6.6 Treatment of steps

Where space on the drawing allows, an arrow pointing up a flight of steps should be used to support level information. The symbol should extend the full length of the flight and must be labelled 'up'.

6.6.7 Depiction of buildings and walls

For roofed structures, the cut line is to be at:

    (a) ground level; or

    (b) sill height; or

    (c) other (specify),

and should show returns for doors and windows on the outside only.

Roofless or ruined structures must have their internal layout (such as walls or columns) shown.

Spot levels must be shown on sills, thresholds and floors.

Annotation indicating floor, wall and roof material as well as building height is to be included.

On plans at scales of 1:100 or larger, floor detail will be required if visible.

Free-standing walls must be shown at a nominal plan height, with lines closed to show openings, where possible.

Additional detail below the plan height (sills, thresholds and floor treatments etc) will either:

    (a) be shown; or

    (b) not be required.

At 1:200 scale detail such as plinths may be omitted if the projection from the wall line is less than 2mm at plot scale.

Where a wall is leaning over significantly from the line of its base, it will be necessary to show the true plan position of both the top (or nominal plan height) and bottom of the wall.

6.6.8 #Above-ground utilities and boundaries

Select the option(s) required.

6.6.8 Above-ground utilities and boundaries

Services, roads, tracks, watercourses, fences, boundaries etc are to be delineated by use of:

(a) surveyed lines; and/or

(b) symbol; and/or

(c) text.

Fence lines are to be indicated by the plotted plan position of posts; the position and width of gates is to be to scale. At 1:500 scale or smaller, building openings, gates and the position of fence posts in plan may be generalised, ie depicted by a symbol or line type.

Ditches are to be shown by a dashed line showing the top of bank. Bottom of bank is to be supported by a spot level at changes of height for each surveyed line.

Overhead services such as telephone or electricity cables are to be shown with a distinctive line type and annotated with the service description and height above OS datum.

Appendix 6.1 CAD layer names for topographic survey

Appendix 6.1 CAD layer names for topographic survey

CAD layer	colour	line type	line weight (mm)		description of content
	Numbers in brackets are AutoCAD colours		up to 1:200	1:500 and over	
0T-breakline	black	dashed	0.18	0.13	in support of contour or hachure, to describe eg top and bottom of slope
0T-cntltxt	black	continuous	0.25	0.18	schedule of coordinates for control stations; to be shown to three decimal places with a description of the marker used; may be included on data sheet for project or as separate file
0T-contour	green	continuous	0.18	0.13	minor contours
0T-contour_index	red	continuous	0.25	0.25	index contours; to be broken to accept contour value; text to be positioned so that the top of the text faces up slope
0T-cut	black	continuous	0.5	0.35	the line of cut for plans
0T-cutP	black	continuous	0.5	0.35	the line of cut for sections and profiles
0T-detail	black	continuous	0.18	0.13	lines used to plot hard detail
	black	continuous	0.18	0.13	wall tops
	black	continuous	0.18	0.13	internal features in roofless buildings
	black	continuous	0.18	0.13	dwarf walls under 300mm high, dashed where edge is uncertain
	black	dashed	0.18	0.13	indicate the position of a wall visible as a sub-contour feature
	black	dashed	0.18	0.13	roof overhangs or buttresses, walls leaning outside wall base
	black	dashed	0.25	0.25	roof overhangs or buttresses, walls leaning inside wall base
0T-footprint	black	continuous	0.25	0.25	ground line, line at the base of a batter or where height of ground becomes part of the building plan
0T-grid	black	continuous	0.25	0.25	Indicate the grid using annotated margin marks and associated text (as per txt layer). Length of line not to exceed 5mm at plot scale. Grid intersections should be shown by an 8mm cross.
0T-gridtxt	black	continuous	0.25	0.18	all text associated with the grid; annotation to be aligned with grid line
0T-hachure	black	continuous	0.18	0.18	hachures
0T-hdge	brown (36)	continuous	0.18	0.13	hedge line at ground level
0T-hgdeOL	light brown (34)	continuous	0.18	0.13	outline limit of hedge spread

0T-inst_cntl	black	continuous	0.25	0.25	all control data with the exception of traverse lines (on layer 0T-trav) and text other than station symbol and target descriptors; datum lines indicated as a 5mm horizontal line on either side of the plotted subject with annotation in text 3mm high; plumb lines to be indicated in the same manner; station symbol to be a triangle with centre mark 3mm high
0T-level	black	continuous	0.25	0.18	level point descriptor
0T-level_text	black	continuous	0.25	0.18	spot levels to two decimal places with associated text rotated so that it is legible with all layers on. Where available space forces the level or any other text to cross other lines a break should be used to ensure clarity.
0T-overhead	black	dashed	0.25	0.18	lines indicating overhead detail
0T-path	black	dashed	0.18	0.13	to show a pathway where there is no kerb or channel
0T-remote	black	continuous	0.25	0.18	lines indicating information remote from the line of cut
0T-subt	black	dashed	0.25	0.18	to show an underground feature such as the path of a traced water course or ice house
0T-surface	black	dashed – dashes to be 2mm with a 1mm gap at plot scale	0.18	0.13	to delineate the outline areas of different ground treatment or material.
0T-svs_d	black	continuous	0.18	0.18	derived service information
0T-svs_elec	red	continuous	0.18	0.18	electrical services – to be expanded as required
0T-svs_fire	red	continuous	0.18	0.18	fire control services eg hydrants
0T-svs_foul	light brown (34)	continuous	0.18	0.18	drainage – foul; show direction of flow
0T-svs_gas	blue	continuous	0.18	0.18	gas services
0T-svs_rw	lilac (175)	continuous	0.18	0.18	drainage – surface water; show direction of flow
0T-svs_tele	orange	continuous	0.18	0.18	for telephone lines; poles to be shown, lines to be shown as an overhead detail with a dashed line
0T-svs_water	turquoise (121)	continuous	0.18	0.18	water supply

0T-text	black	continuous	0.25	0.18	all text except title, control text, tree data and text associated with height information. Text should be positioned to avoid overwriting detail when the layer is on with all other layers.
0T-title	black	continuous	0.35	0.25	rubric, key, logos, north signs, scale bars and all associated text
0T-trav	black	continuous	0.25	0.18	traverse lines with annotation of reduced angles, distances and station coordinates (if other than WORLD coordinate system is used)
0T-tree	green	continuous	0.25	0.25	tree bole hatched solid
0T-treeA	green	continuous	0.18	0.13	canopy spread by individual tree
0T-treeB	green	continuous	0.18	0.13	canopy spread trimmed to envelope
0T-treeM	green	continuous	0.18	0.13	tree less than 5m in height
0T-treetxt	green	continuous	0.25	0.18	tree – descriptive text
0T-veg1	dark green (96)	continuous	0.18	0.13	limit of vegetation, to be subdivided if needed
0T-wall_top	black	continuous	0.25	0.25	lines used to describe wall tops inside the line of cut (ie the view of the wall looking from above) if this varies significantly from the cutline

This is not an exhaustive list. New layers may be created so long as they are prefixed with 0T-.

(a) General considerations

The cut line of a building or feature should be of a heavier weight than lines used for other detail.

A dot and peck line type should be used to indicate any or all of the line types in the table if there is a conflict of lines and for boundaries if required to avoid confusion.

A dotted line may be used for clarity if there are a large number of dashed lines on the drawing sheet.

In AutoCAD, LTGEN is to be set to on.

(b) Dashed lines

The line type should be controlled so that dashes are 0.5mm long with a 0.5mm gap at the plot scale. The exception is for lines showing changes in surface treatment where the dashes should be 2mm with a 1mm gap.

(c) Dotted lines

A dotted line should be a 0.18mm or 0.25mm diameter dot at a 2mm to 5mm interval, depending on the map scale plotted.

(d) Dot and peck lines

The dot and peck line should comprise a line 1mm in length separated by a 0.5mm gap from a dot of 0.18mm, with a 0.18mm line width.

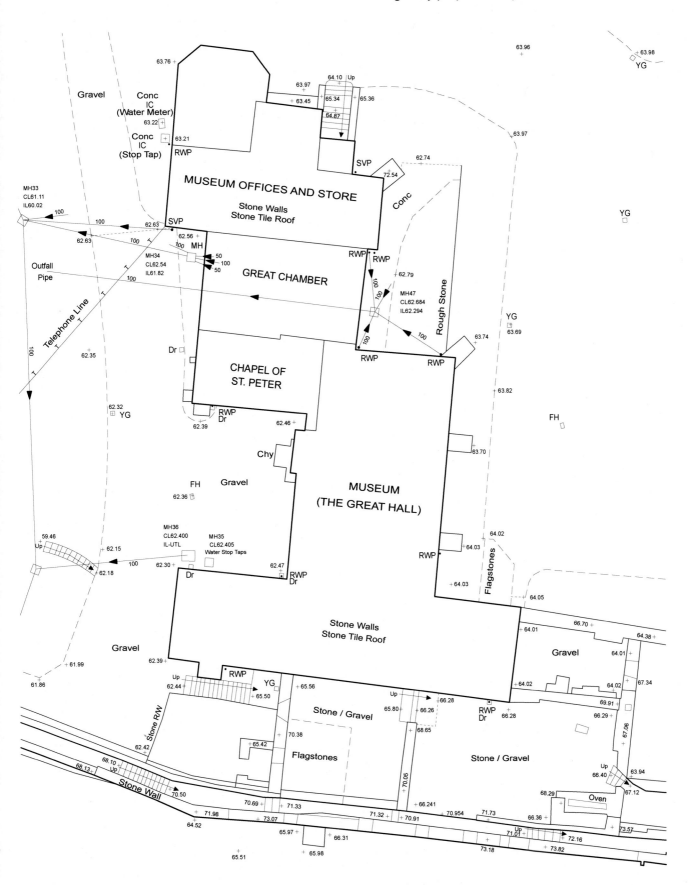

Fig 6.1 Occupied buildings and associated detail.
Originally prepared for presentation at 1:200 scale

Fig 6.2 Depiction of trees; top, canopies untrimmed; bottom canopies trimmed to the overall envelope

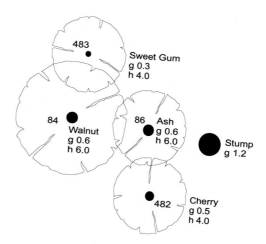

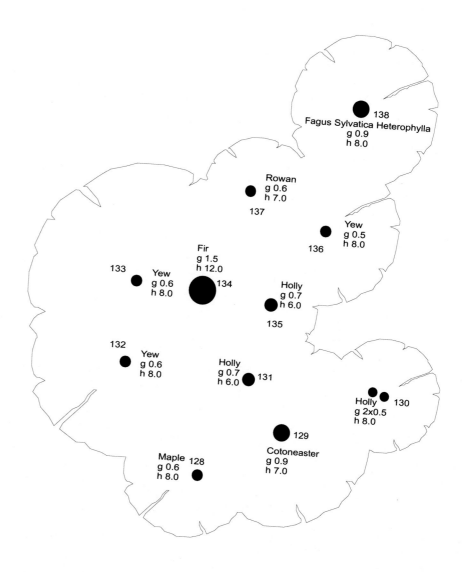

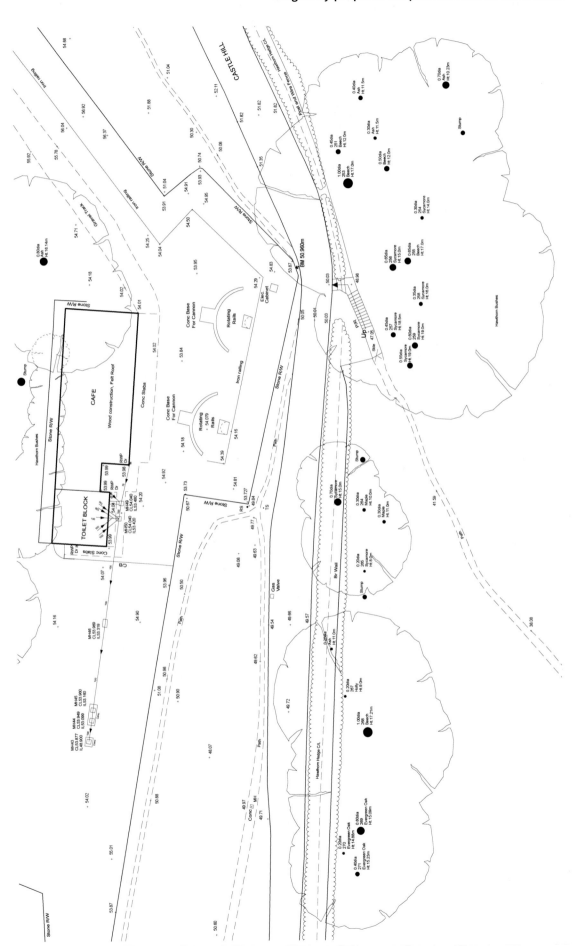

Fig 6.3 Depiction of hedges (including centreline).
Originally prepared for presentation at 1:200 scale

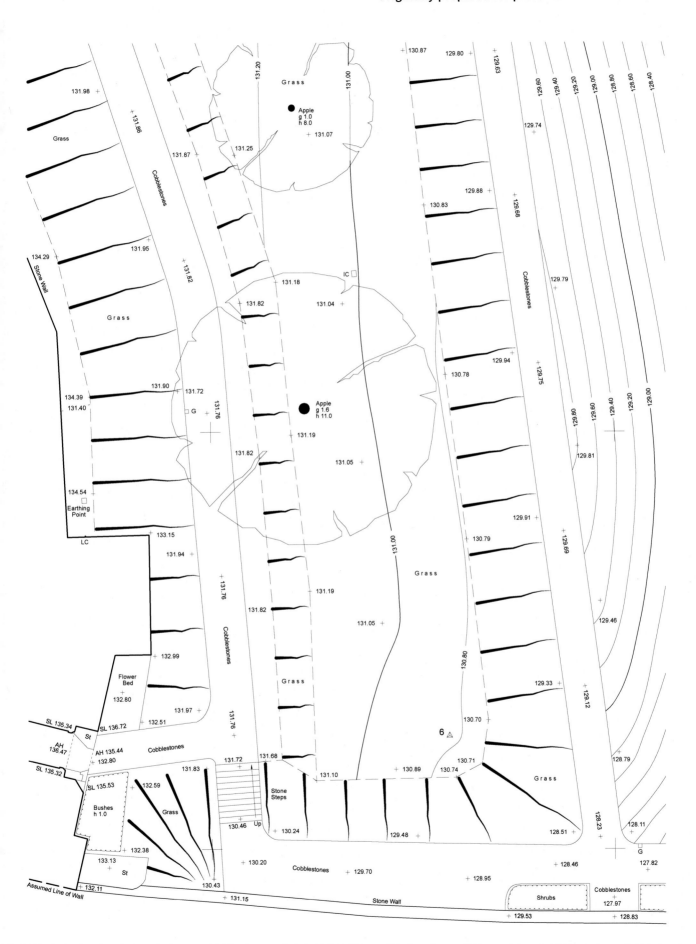

Fig 6.4 Use of hachures to supplement information from contours.
Originally prepared for presentation at 1:200 scale

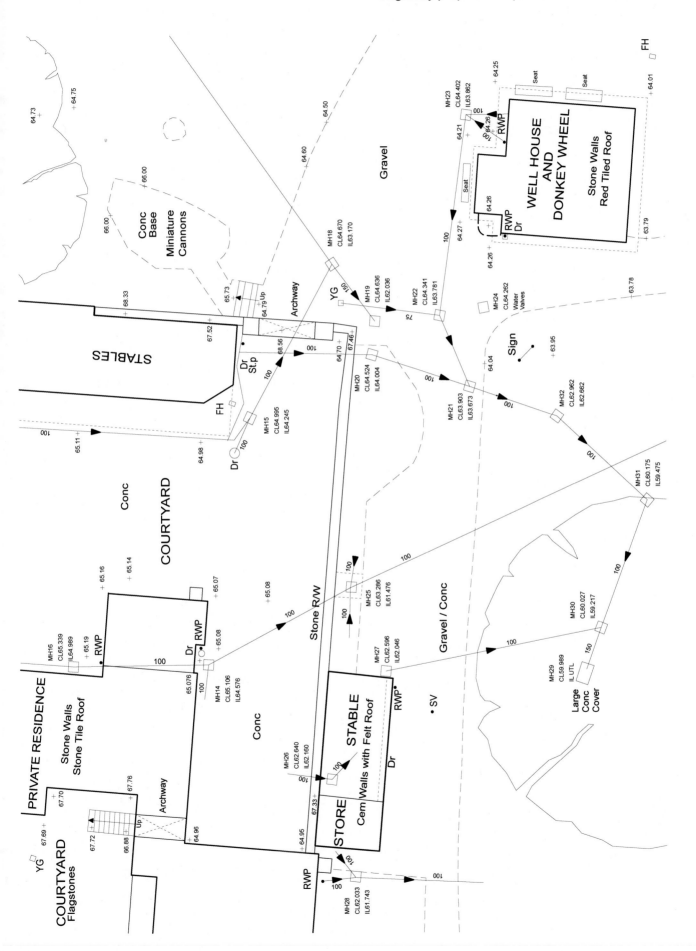

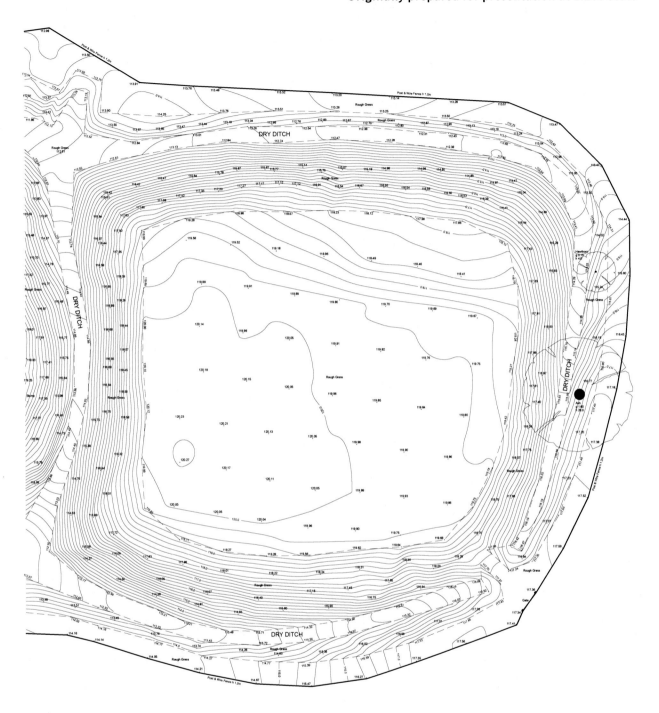

Section 7

Standard specification for the collection, registration and archiving of laser scan data

7.1 Definitions of terms

7.1.1 Laser scanning

The term 'laser scanner' is generally applied to a range of instruments that operate with differing measurement principles, in different environments and with different levels of precision and resulting accuracy. A laser scanner can be defined as 'any device that collects 3-D coordinates of a given region of an object's surface automatically and in a systematic pattern at a high rate achieving the results in near real time' (adapted from Böhler, W and Marbs, A 2002 '3D Scanning Instruments'. Proceedings of CIPA WG6 Scanning for Cultural Heritage Recording, September 1–2, Corfu, Greece). There are three main types of laser scanner:

- Close-range: using triangulation or structured light technology, they are more suited to artefact-sized objects.

- Terrestrial: using time of flight and/or phase comparison technology, they are generally used for building-sized objects.

- Aerial: using lidar (light detection and ranging) technology, they are more suited to landscape-sized subjects and will not be covered by this document. For an in-depth guide to aerial laser scanning, *see* the English Heritage publication, *The Light Fantastic: Using airborne lidar in archaeological survey* (Crutchley, S and Crow, P 2009), available for free download from **www.historicengland.org.uk/images-books/publications/light-fantastic/**.

7.1.2 Point cloud

Point clouds may be captured for a wide variety of heritage subjects including artefacts, architectural details, buildings, monuments and entire historic landscapes. They are ideal for the collection of three-dimensional surface information from which other outputs can be extracted. They do not, however, easily lend themselves to the automatic generation of some survey products, such as line drawings, that are traditionally used in cultural heritage applications. As with photogrammetry, a skilled operator is required to extract drawing information. In some circumstances a laser scanner will be the most appropriate tool, such as when lack of suitable texture or lighting inhibits the use of photogrammetry. Point clouds should not normally be seen as a replacement for existing survey products or as an end product in their own right.

7.1.3 Point density

Point density is the average distance between XYZ coordinates in a point cloud. It is commonly represented in two forms:

Spatial – the average distance between points at a specified range eg 6mm at 10m

Angular – the angular step of the two axes – eg 0.25° × 0.25°

Historic England

Metric Survey Specifications for Cultural Heritage | **Section 7: Standard specification for the collection, registration and archiving of laser scan data**

7.1 Definitions of terms

7.1.1 Laser scanning

Terrestrial laser scanning is defined as the use of a ground based device that employs a laser to automatically measure three-dimensional coordinates on the surface of an object in a systematic order and at a high measurement rate. For the purposes of this document close-range and terrestrial laser scanners will jointly be referred to as terrestrial laser scanners.

7.1.2 Point cloud

Any laser scanning system generates a point cloud which can be regarded as the raw product of a laser scan survey. A point cloud is a collection of XYZ coordinates in a common coordinate system that portrays to the viewer an understanding of the spatial distribution of a subject. It may also include additional information, such as return intensity or even colour values. Generally, a point cloud contains a relatively large number of coordinates in comparison with the volume the cloud occupies, rather than a few widely distributed points. Some instruments also provide more fundamental information on the full reflectance of the laser pulse (known as full-waveform scanners).

7.1.3 Point density

This is the average distance between XYZ coordinates in a point cloud represented in a spatial or angular manner.

Where a spatial value is given, this must be supported by the suggested range from the scanner to the subject. The quoted value should always be given as the minimum density of points required to suitably capture and represent the detail on the object.

7.2 Data collection

7.2.1 Pre-survey deliverables

To ensure clarity in the requirements of the survey a number of pre-survey deliverables are required prior to laser scanning being undertaken on site. This can take the form of a method statement that, in the case of laser scanning, should at a minimum include:

- technical specifications of the proposed scanning system(s);

- the proposed point density of the scans;

- the proposed targeting and registration approach; and

- the location of potential data voids and how they might be in filled.

7.2.2 #Certification requirements

Choose an option. Although there is currently no industry-wide standard for the certification of terrestrial laser scanners, evidence of the proposed laser scanner's calibration should be requested prior to work commencing on site.

7.2.3 Health and safety

Laser light can, in some cases, be harmful and therefore suitable precautions must be taken to ensure the safety of both the scanner operator and any members of the public that may encroach within the area of scanning.

Please *see* Appendix 1.5 for further details and BS EN 60825-1: 2014 for full details of the precautions required in the use of lasers.

Historic England

Metric Survey Specifications for Cultural Heritage | **Section 7: Standard specification for the collection, registration and archiving of laser scan data**

7.2 Data collection

7.2.1 Pre-survey deliverables

A method statement, as defined in section 1.3.2, must be supplied prior to survey being undertaken on site.

7.2.2 Certification requirements

The method statement is to be accompanied by either:

(a) a manufacturer supplied calibration certificate for the proposed laser scanner; or

(b) details of tests performed within the last 12 months that demonstrate the geometric precision and accuracy of data captured by the laser scanner.

7.2.3 Health and safety

Systems that use Class 1, 2 and 3R lasers operate within acceptable eye-safe limits and therefore can be used for survey purposes at an English Heritage site.

Systems that use Class 3B or 4 lasers operate beyond recognised eye-safe limits and therefore must *not* be used for any survey-related activity at an English Heritage site.

Signs warning visitors that scanning lasers are in use should be displayed in each scan area. These are mandatory for Class 2 and 3R laser scanning systems and optional for those using Class 1 lasers.

Historic England

Metric Survey Specifications for Cultural Heritage | **Section 7: Standard specification for the collection, registration and archiving of laser scan data**

7.2.4 #Point density and measurement precision

Choose an option. The point density of the required 360° scans should be chosen with reference to both the accuracy of individual points measured by the chosen scanner and the required accuracy of output. Scanning at a higher density than the accuracy of the measurement may generate an impressive dataset but will result in a high level of noise within the resulting point cloud.

The smaller the distance between points, the more likely it is that an object will be recorded. So as a general rule the point density should be at least half the size of the smallest feature to be recorded within the scan.

In order to minimise the creation of scanning artefacts within the scanning process, point densities should be equal in both scanning axes. For example, uneven point densities may lead to the lack of definition of horizontal or vertical features. Also, the beam width of the measurement beam must not be greater than double the effective point density.

For most instruments point density during the scanning process depends on range, so it is not normally possible to maintain a constant point density over the entire subject. It is most likely, however, that a particular area will be of interest, such as a facade or building detail, where a regular density of points is preferable.

7.2.5 #Window scans

Choose an option. Although the majority of scanning projects are undertaken using 360° panoramic scans, there may be a need to scan specific areas at a higher density in order to create either a better quality or a different type of output, eg a specific architectural feature or area of sculptured detail upon an elevation. This is typically achieved using window scans, where the area of interest is delineated on a previously acquired 360° scan and the specific resolution requirements set on the scanner prior to acquisition. If window scans are needed ensure that the required areas are clearly communicated by means of a plan or detailed description.

7.2.6 #Overlapping scans

Choose an option. Overlapping scans must be acquired to ensure that a full record of a subject is collected as well as providing users with confirmation that the registration process has been successful.

Once registered it is possible to filter overlapping scan data in order to reduce the point density in the final registered point cloud and hence reduce file sizes and demands on software systems during processing. This is generally not recommended in order to retain the original density of the raw scans and ensure that a full archive of the subject is collected.

Historic England

Metric Survey Specifications for Cultural Heritage | **Section 7: Standard specification for the collection, registration and archiving of laser scan data**

7.2.4 Point density and measurement precision

The required point density and precision is to be either:

(a) as defined below for the scale specified in the project brief (1.1.10); or

	scale	effective point density	precision of measurement
Close-range	1:5	0.5mm	+/- 0.5mm
	1:10	1.0mm	+/- 1.0mm
Terrestrial	1:20	2.5mm	+/- 2.5mm
	1:50	5.0mm	+/- 5.0mm
	1:100	15.0mm	+/- 15.0mm

(b) based on a minimum feature size of ......mm.

7.2.5 Window scans

The use of window scans to record specific areas at a higher resolution is either:

(a) required at a point density of ......mm (the specific areas are highlighted on the attached plan); or

(b) not required.

7.2.6 Overlapping scans

Filtering areas of overlapping scan data, so as to reduce the point density in the final registered point cloud, is either:

(a) required; or

(b) not required.

Historic England

Metric Survey Specifications for Cultural Heritage | Section 7: Standard specification for the collection, registration and archiving of laser scan data

7.2.7 Data voids

Data collected during the survey of cultural heritage sites will form part of their archival record so data voids should be kept to an absolute minimum. Although site constraints may mean data voids are unavoidable they can be minimised through careful selection of overlapping scan positions. Temporary obstructions to the scanner during observation, such as those caused by passing vehicles and pedestrians, should also be minimised through careful control of public access into the scan area.

7.2.8 High level coverage

Laser scanning relies on line of sight observation, but where a subject is of significant height locating a suitably high vantage point, adjacent to the scan area, may prove difficult. Suitably high scanning platforms may be temporarily provided by a scaffold tower, hydraulic lift, mast or even small unmanned aircraft (SUA). However, as laser scanning is a continuous process that typically takes minutes to complete, any slight movement in the platform will generate movement of the scanner head and inaccuracy in the resultant point cloud. Therefore where temporary solutions are suggested the effect of high-level movement on the scanner must be considered and suitably controlled.

7.2.9 Weather

Some forms of weather will impact on the quality of data collected, either by introducing noise into the point cloud or simply by affecting operation of the scanner itself. For example, scanning in heavy rain or snow will lead to erroneous data returns from airborne water droplets and, if left to accumulate on the scanner mirror, incorrect range measurements due to refraction of the measurement beam.

7.2.10 #Control and coordinate systems

Choose an option. It is usual practice to georeference terrestrial point cloud data to a known local or national coordinate system either by GNSS measurements to scan control targets or by observations to existing ground control points. If such a site coordinate system already exists, full details, including coordinate listing and witness diagrams, must be supplied to the scanning operator/contractor to enable reoccupation. Where a previous coordinate system does not exist, a new system may need to be established; *see* section 2.2 for further details.

Historic England

Metric Survey Specifications for Cultural Heritage | **Section 7: Standard specification for the collection, registration and archiving of laser scan data**

7.2.7 Data voids

Voids caused by temporary obstructions, such as cars and pedestrians, will not be acceptable unless it is impossible to restrict public access into the scan area. Any obstructions that will potentially occlude areas in the scan should be highlighted in the method statement. Alternative measurement technologies for potentially infilling the data void may also be proposed.

7.2.8 High level coverage

Methods used to achieve high level coverage must be described in the method statement alongside any proposed mechanisms for stabilising the platform and laser scanner.

7.2.9 Weather

When working outdoors, the weather conditions during the survey should be recorded, although where possible scanning in extremes of weather is to be avoided. If not practicable, suitable protection for the scanner must be employed to minimise its effect.

7.2.10 Control and coordinate systems

The methods and networks used for providing survey control are discretionary. However, details of the proposed method and equipment must be included in the method statement.

Georeferencing of point cloud data is to be facilitated either:

(a) by observation to existing ground control points; or

(b) through the establishment of a new control system,

as described in section 2.2.

Historic England

Metric Survey Specifications for Cultural Heritage | Section 7: Standard specification for the collection, registration and archiving of laser scan data

7.2.11 #Targets and registration

Choose an option or allow contractor's discretion. Although a single scan may be sufficient to fully record certain scenes, multiple scans are likely to be required, especially when dealing with a large site or structure. Targets located in a defined coordinate system will usually be used to register such multiple scans together. Positioned using a TST or GNSS receiver, they provide additional checks on the geometric quality of the scan data and enable the transformation of the complete dataset to a common coordinate system. The design of targets does vary between laser scanner manufacturers, but they are usually spherical, hemispherical or planar in form, with a white and/or grey surface to maximise laser returns.

Although natural points of detail may also be used as control points, these should generally be avoided, as scan artefacts may occur at geometric edges where the footprint of the laser beam hits multiple surfaces, resulting in erroneous points being recorded.

It is possible to register laser scan data together without the use of any external reference targets. Known as 'cloud-to-cloud', this target-less registration approach relies on suitably dense point clouds being captured with appropriate overlap between adjacent scans to enable sufficient matching points to be derived. Although this approach typically requires more scans to be captured, it can result in registration statistics similar to those achieved using target-based approaches. It is also a useful approach when working in environments, where control targets are at risk of being accidentally moved, or when wishing to achieve a point cloud without any visible reference targets.

Historic England

Metric Survey Specifications for Cultural Heritage | **Section 7: Standard specification for the collection, registration and archiving of laser scan data**

7.2.11 Targets and registration

Scan registration must be performed using either:

(a) control targets placed within each scan; and/or

(b) a target-less 'cloud-to-cloud' approach.

The type of targets that may be used, their positioning around a site and the method of laser scan observation are all discretionary. They must, however, be:

- not so large that they obscure important fabric detail;

- positioned away from the principal surface being surveyed;

- not attached to any important historic fabric;

- arranged so as to minimise data voids in the point cloud; and

- removed at the end of the survey, either on completion of site work or after successful registration is achieved.

A description of the targets and proposed registration approaches must be included in the method statement.

7.3 Provision of point cloud data

7.3.1 #Deliverables

Complete the table as appropriate. The range of deliverables that that can be derived from a laser scan survey will vary from project to project. However, to maximise their application potential and long-term archival value it is suggested that as a minimum the raw and registered scan data, individual scan metadata and associated survey report are requested, alongside any specified outputs derived from the scanning.

There is an increasing application of web-based approaches to the distribution of laser scan datasets without the need to use or access high-end processing software. Once the required format has been created through the post-processing software operation, free web-based tools can be downloaded from some manufacturers' websites that enable viewing, manipulation and measurement extraction using either a server, hard-drive or cloud-based application. Examples of such solutions include Leica TruView and FARO WebShare.

Historic England

Metric Survey Specifications for Cultural Heritage | **Section 7: Standard specification for the collection, registration and archiving of laser scan data**

7.3 Provision of point cloud data

7.3.1 Deliverables

The following deliverables are required (*see* sections 3.4 and 7.4 for a full description of the appropriate media, formats and required metadata).

Deliverables		Required	Number of copies	Not required
Raw scan data	Proprietary format			
	Non-proprietary format			
Registered scan data	Proprietary format			
	Non-proprietary format			
Metadata	For individual scans			
Survey report	For whole project			
Derived outputs (*see* sections 4, 5 and 6)	Drawings			
	Orthorectified imagery			
	3-D surface models			
Web-based data	Specify format			

The survey report is to be supplied as a PDF file and as a minimum must include:

- a diagram showing the approximate scan positions;

- a diagram showing the location of all targets;

- the traverse/control network diagram;

- a listing of 3-D coordinates of all control points/targets; and

- a registration report showing the overall accuracy of the laser scan survey.

Historic England

Metric Survey Specifications for Cultural Heritage | **Section 7: Standard specification for the collection, registration and archiving of laser scan data**

7.3.2 #Intensity/colour information

Choose an option. All laser scanning systems provide intensity information alongside the XYZ coordinates for each 3-D point that comprises the point cloud. Many also use an imaging sensor, either located within the scanner or attached externally, to acquire separate colour imagery of the scan scene from which an RGB value for each point can be extracted. Depending on the scanner the capture of additional imagery can add substantially to the scanning times on site, though it does enable the colourisation of point cloud data and the subsequent creation of colour outputs.

7.3.3 #Additional image data

Choose the required option(s). Separate digital imagery taken of both the subject being scanned and its surrounding context can prove useful within later post-processing operations. As well as providing visual validation of the 3-D data recorded by the scanner, it can, if taken at an appropriately high resolution and from suitable positions, be used as complementary content during later post-processing and aid the generation of higher-quality, image-based outputs.

Historic England

Metric Survey Specifications for Cultural Heritage | Section 7: Standard specification for the collection, registration and archiving of laser scan data

7.3.2 Intensity/colour information

Alongside the laser intensity value, RGB colour information, acquired on a per point basis at each scan position, is either:

(a) required; or

(b) not required.

7.3.3 Additional image data

Additional digital imagery, showing the subject being scanned and the surrounding context, is either:

(a) required; or

(b) not required.

Panoramic imagery acquired with a separate camera, the optical centre of which has been aligned with the laser scanner measurement centre, is either:

(a) required; or

(b) not required.

All imagery is to be delivered as specified in section 3.1.

Historic England

Metric Survey Specifications for Cultural Heritage | **Section 7: Standard specification for the collection, registration and archiving of laser scan data**

7.4 Storage and archive of point cloud data

7.4.1 #Data format

Choose an option. To assist in the future management of laser scan datasets, all data must be delivered in pre-specified formats with emphasis on both the usability and transferability of data between different software systems.

The raw scan data (as collected by the scanner before processing) and any later registered versions should be delivered in both its proprietary (manufacturer specific) format and in the non-proprietary E57 data exchange format recently developed by the American Society for Testing and Materials (ASTM). This is a format for storing point clouds, images and metadata produced by 3-D imaging systems. Further details of the E57 format can be found at the ASTM website (**www.astm.org/Standards/E2807.htm**), as well as through the individual laser scanner manufacturers, who now include it as both an import and export option within their post-processing software.

7.4.2 #File naming system

Choose an option. Filenames for laser scan datasets should all conform to the standard naming system used throughout this specification – *see* section 3.1.1 for further details. However, there may be applications where a non-standard approach may be justified as long as it observes the following rules:

- Use only alphanumeric characters (a–z, 0–9), the hyphen (-) and the underscore (_).

- Both upper and lower case characters and numbers may be used.

- Spaces and full stops (.) must not be used within filenames – a full stop must only be used to separate the filename from the extension.

- Descriptive filenames may be used.

- Non-descriptive filenames are acceptable but their content must be adequately described in accompanying metadata.

Historic England

Metric Survey Specifications for Cultural Heritage | **Section 7: Standard specification for the collection, registration and archiving of laser scan data**

7.4 Storage and archive of point cloud data

7.4.1 Data format

Both the raw and registered versions of the data are to be delivered in either:

   (a) the scanner proprietary and E57 data exchange format; and/or

   (b) other format (specify).

7.4.2 File naming system

All laser scan files are to be named using either:

   (a) the file naming system as specified in section 3.1.1; or

   (b) the contractor proposed convention (to be set out in the method statement).

Historic England

Metric Survey Specifications for Cultural Heritage | **Section 7: Standard specification for the collection, registration and archiving of laser scan data**

7.4.3 Scan metadata

In order to achieve the best records for digital dissemination and archiving, a suitable level of digital documentation – often referred to as metadata or 'data about your data' – must be collated. It should explain what was done and why, as well as where and when it was undertaken. At its simplest, metadata should provide the information that someone would require to successfully regenerate your digital records from the files deposited in the archive.

> English Heritage, 2006 *MoRPHE Technical Guidance 1: Digital Archiving and Digital Dissemination.*

Example:

parameter	example
raw data file name:	KEN14L01.FLS
project reference number:	Not applicable
scanning system used and serial number:	Faro Focus 3D S 120 #LLS061101896
average point density on the subject:	0.006 m (@ 10 m)
total number of points:	857,446
date of capture:	08/07/2014
site name:	Kenilworth Castle
list entry number:	1014041
company name:	English Heritage

Historic England

Metric Survey Specifications for Cultural Heritage | Section 7: Standard specification for the collection, registration and archiving of laser scan data

7.4.3 Scan metadata

Metadata (information relating to the captured scan data) must be supplied with all raw scan data and as a minimum is to include the following:

- raw data file name;

- project reference number (if known);

- scanning system used including serial number;

- average point density on the subject (with reference range);

- total number of points;

- date of capture;

- site name;

- list entry number (if known); and

- company name.

This metadata can either be provided in digital form for each individual scan or incorporated within each scan data file, ensuring that the required data fields are correctly included and can be retrieved using the common post-processing software.

Project-level metadata is to be provided in the form of the survey report, ensuring the items noted in section 7.3.1 are included.

Historic England

Metric Survey Specifications for Cultural Heritage | Section 7: Standard specification for the collection, registration and archiving of laser scan data

Section 8

Standard specification for the supply of building information modelling (BIM)

8.1 Definitions of terms

8.1.1 What is BIM?

Building information modelling (BIM) is a digital tool to aid better information management, which combines the benefits of 3-D geometric modelling of a building with an accurate understanding of the architectural components from which it is constructed.

The UK Government Construction Strategy was published by the Cabinet Office on 31 May 2011, and although the report refers to different levels and associated timescales for BIM implementation, it announced the government's intention to require collaborative 3D BIM at level 2 (with all project and asset information, documentation and data being electronic) for its projects by 2016 (BIM Task Group 2014).

The levels referred to are as follows:

- *Level 0:* unmanaged 2-D CAD data, typically in paper form, that has been used within a heritage context since the late 1980s;

- *Level 1:* managed 2-D and 3-D CAD data, typically in electronic form using a common data environment and to defined standards;

- *Level 2:* managed 3-D information in electronic form that defines the physical and functional characteristics of a building; and

- *Level 3:* a single integrated (iBIM) model that includes and accesses all available data forms for the project.

Fig 8.1 The development of BIM and its usage (after Bew and Richards)

BIM Maturity Diagram

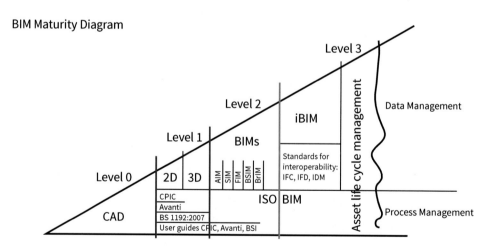

8.1.2 How can BIM be used within a heritage project?

The UK government's promotion of BIM has meant that the current focus and application is principally within new-build design and construction. However, the development of geometric modelling with defined component information also provides potential scope for application to existing historic buildings within their recording, architectural/archaeological assessment, conservation planning, presentation and long-term management.

Historic England

Metric Survey Specifications for Cultural Heritage | **Section 8: Standard specification for the supply of building information modelling (BIM)**

8.1 Definitions of terms

8.1.1 What is BIM?

The term 'BIM' is the widely used acronym for building information modelling, which is the process of digitally representing the physical and functional characteristics of a building.

8.1.2 How can BIM be used within a heritage project?

BIM is a process that enables the entire project team to collaborate and create a single source of data that assists design, construction, facilities management and other related processes by allowing the extraction and updating of information throughout the life cycle of the building.

Historic England

Metric Survey Specifications for Cultural Heritage | **Section 8: Standard specification for the supply of building information modelling (BIM)**

8.1.3 Purpose and scope of this document

The BIM process for a building typically requires on-site acquisition of an appropriate level of geospatial data, followed by its use in creating the required 3-D geometric models, with associated component information. Commonly referred to as 'BIM-ready' this preliminary dataset can then be supplied to other relevant parties and additional datasets/links inserted that provide a single source of BIM data that facilitates collaboration and long-term management of the building.

Historic England

Metric Survey Specifications for Cultural Heritage | **Section 8: Standard specification for the supply of building information modelling (BIM)**

8.1.3 Purpose and scope of this document

This section of the specification relates to the creation of a 'BIM-ready' dataset. The initial capture of the required level of geospatial data will be defined in sections 2, 4, 5, 6 and 7 of this specification as appropriate.

Historic England

Metric Survey Specifications for Cultural Heritage | **Section 8: Standard specification for the supply of building information modelling (BIM)**

8.2 BIM development

8.2.1 #Level of detail

Choose the required level of detail and note any variations required.

The BIM process allows the production of 3-D geometric models typically based on previously acquired three-dimensional datasets, as supplied by laser scanning and photogrammetry, although other less complex, two-dimensional survey sources can also be used. Each modelled component contains additional information about the building fabric employed during construction which allows digital representation of the structure and how the components fit together.

Buildings, whether modern or historic, typically share similar architectural features which can be represented in a BIM environment by generic, 'off the shelf' components. However, for accurate representation and to allow future interrogation, analysis and presentation opportunities, additional geometric detail may need to be modelled for each component. Accurate architectural modelling takes time and leads to larger, more complex datasets that may prove difficult to use. Therefore four levels of detail are suggested when digitally representing a building in a BIM.

8.2.2 #Modelling tolerance

Within historic BIM applications the current process attempts to fit irregular, real-world features into a regular, software-based environment. However, accurate architectural modelling takes time and complete understanding of the building requires in-depth knowledge of the fabric. Consequently, to provide a product that is realistic in terms of both time and cost often demands an acceptance of specific tolerances being applied within the modelling process. Once created a BIM model does, however, enable the physical and functional characteristics of a building to be determined both with regard to how the components fit together and how it performs as a working structure.

In general the tolerances will be defined by the level of detail. Any required variation, for example for particular components, should be tabulated in this section.

Historic England

Metric Survey Specifications for Cultural Heritage | **Section 8: Standard specification for the supply of building information modelling (BIM)**

8.2 BIM development

8.2.1 Level of detail

The required BIM data is to be constructed to the following level of detail:

(a) *Level 1:* basic outline of the building/structure represented as a solid object using representative component information but with no architectural detail depicted; or

(b) *Level 2:* outline of the building/structure represented as a solid object with principal architectural features included using generic components; or

(c) *Level 3:* outline of the building/structure represented as a solid object with all architectural features and major service detail included using generic components; or

(d) *Level 4:* detailed survey of the building/structure represented as a solid object including all architectural detail, services and custom developed components to accurately represent fabric type; or

(e) other (specify).

See Fig 8.2 for example of levels 2 to 4.

Specific features to be modelled or custom components to be developed are tabulated below......

8.2.2 Modelling tolerance

The generally accepted tolerances of BIM modelling are as defined by the level of detail selection.

Variations required for individual components, are tabulated below......

Historic England

Metric Survey Specifications for Cultural Heritage | **Section 8: Standard specification for the supply of building information modelling (BIM)**

8.3 Supply of BIM data

8.3.1 #BIM data formats

Choose an option. For the purposes of this specification two data formats may be used for the exchange of BIM data – IFC and RVT.

The Industry Foundation Classes (IFC) data model is an industry-wide, open and neutral data format. Although file sizes are typically larger than the original model, it is fast becoming the standard for non-proprietary data exchange of BIM information due to its interoperability between different software platforms. However, it is important to use an up-to-date version, to minimise data errors that may be encountered during import. For this reason it is recommended that IFC versions 2.0 and above are used.

RVT is a proprietary data format used by Autodesk within their Revit Architecture software, part of the Building Design suite. It is the native format of Autodesk Revit and so will provide both the smallest file size and easiest exchange of data across Autodesk platforms. However, it is only forward-compatible, so it is recommended that RVT versions 2014 and above are used although, to minimise data errors that may be encountered during importation, later versions are preferable.

8.3.2 #Deliverables

Choose the required options. Alongside the BIM data itself it will be useful to receive the component library from which it is constructed, particularly where custom-developed objects have been created. Also, to increase the potential for physical analysis and provide a real-time comparison of how the modelled components fit the original survey data, the 3-D files upon which they are based should be supplied. Ideally these will be spatially registered with the BIM data.

Historic England

Metric Survey Specifications for Cultural Heritage | **Section 8: Standard specification for the supply of building information modelling (BIM)**

8.3 Supply of BIM data

8.3.1 BIM data formats

All BIM-ready files are to be supplied as:

    (a) IFC version...... ; and/or

    (b) Autodesk Revit version.......RVT; and/or

    (c) other (specify).

8.3.2 Deliverables

The BIM data is to be supplied:

    (a) with the component library; or

    (b) without the component library but referencing that which was used; or

    (c) using another form of library information (specify).

The base 3-D survey files:

    (a) are to be included and registered with the BIM data; or

    (b) are not required; or

    (c) are to be provided in another form (specify).

Historic England

Metric Survey Specifications for Cultural Heritage | **Section 8: Standard specification for the supply of building information modelling (BIM)**

8.4 BIM references

8.4.1 #Existing standards

The last three years have seen rapid growth in BIM implementation across the UK, leading to a wealth of reference material being developed and becoming available to both suppliers and potential end-users.

8.4.2 BIM Task Group

As noted on their website (**www.bimtaskgroup.org**) 'the Building Information Modelling (BIM) Task Group are supporting and helping deliver the objectives of the Government Construction Strategy and the requirement to strengthen the public sector's capability in BIM implementation with the aim that all central government departments will be adopting, as a minimum, collaborative Level 2 BIM by 2016'.

Although the group have not yet developed any BIM-related specifications, by bringing together expertise from industry, government, public sector, institutes and academia they have collated a wealth of knowledge on the full BIM process and therefore serve as an essential resource when considering any application, whether new-build or for an existing heritage structure.

8.4.3 BIM4Conservation (BIM4C)

Established as part of the BIM 4 Communities programme, this group aims to raise awareness and understanding of BIM within the conservation and heritage sector of the built environment and to promote the benefits that BIM brings to the stewardship of existing buildings.

8.4.4 Royal Institute of Chartered Surveyors (RICS)

The professional body that represents everything professional and ethical in land, property and construction, RICS is well placed to support and raise awareness for the BIM agenda within the UK survey industry. Although no generic BIM specifications have yet been developed, it does provide a variety of information on the process through its website (**www.rics.org/bim**), alongside a professional certificate in BIM Implementation and Management.

8.4.5 Royal Institute of British Architects (RIBA)

The professional body that champions better buildings, communities and the environment through architecture and its members. Through its BIM Overlay publication (RIBA, May 2012), part of the RIBA Plan of Work (**www.ribabookshops.com/plan-of-work**), it provides straightforward guidance on the activities needed at each RIBA work stage to successfully design and manage construction projects within a BIM environment.

8.4.6 The Survey Association (TSA)

The trade body for commercial survey companies in the UK that provides client guides and guidance notes on various areas of surveying. Although no TSA guide on BIM currently exists, its Technical Committee is considering BIM alongside other technologies and approaches relevant to surveyors (**www.tsa-uk.org.uk/tag/bim/**).

 Historic England

Metric Survey Specifications for Cultural Heritage | **Section 8: Standard specification for the supply of building information modelling (BIM)**

8.4 BIM references

8.4.1 Existing standards

Where there is a specific requirement to reference an existing BIM standard the title, section, paragraph and relevance to the project will be inserted here.

Historic England

Metric Survey Specifications for Cultural Heritage | **Section 8: Standard specification for the supply of building information modelling (BIM)**

8.4.7 Commercial survey suppliers

A number of commercial survey providers have developed their own BIM survey specifications. As well as defining the capture, modelling and delivery processes that will be employed on BIM projects these also aid in determining initial client requirements and expectations when needing to incorporate the survey of buildings, land and utilities within a BIM environment.

Historic England

Metric Survey Specifications for Cultural Heritage | **Section 8: Standard specification for the supply of building information modelling (BIM)**